101
SOUND, LIGHT AND POWER IC PROJECTS

No. 2604
$24.95

101 SOUND, LIGHT AND POWER IC PROJECTS

CHARLES SHOEMAKER

FIRST EDITION

SECOND PRINTING

Printed in the United States of America

Library of Congress Cataloging in Publication Data

Shoemaker, Charles.
101 sound, light, and power IC projects.

Includes bibliographical references and index.
1. Electronics—Amateurs' manuals. 2. Printed circuits—Amateurs' manuals. I. Title. II. Title: One hundred one sound, light, and power IC projects.
TK9965.S495 1986 621.381 85-27483
ISBN 0-8306-9604-0
ISBN 0-8306-0404-9 (pbk.)

Contents

Introduction

INTEGRATED CIRCUITS ARE IMPORTANT TO THE HOBBYIST, THE EXperimenter, and the student, because they provide the designer with unlimited possibilities for interfacing to accomplish a given function. They can be mounted on proto-boards along with discrete and passive components like transistors, diodes, capacitors, and resistors for study and analysis to accomplish a given purpose. The trend toward more complex integrated circuits will continue, which means greater possibilities in a futuristic technology.

Control through integrated circuits is a way of communicating electronically from humans to humans, from machines to machines, or between humans and machines. Sound, light, and power control are communicating elements to which integrated circuits have wide application. The extension of our ears, eyes, and fingers in the chain of communications is essential for greater human awareness.

This book was written to provide learning where proto-board usage makes circuit adaptation and experimental design a great potential. The proto-board procedure requires high concentration and a disciplined skill of working carefully and accurately. The beginning student should start with the less complex circuits and each circuit should perform as explained before changes in the "Further Study" sections are made.

Chapter 1 identifies important considerations when working with the problems in the book. Chapter 2 describes the process of making an etched circuit board. Chapter 3 explains the usage of the proto-board for designing and building circuits in the first phase of circuit development. Chapter 4 shows the way of making multiple circuit boards for quantity production by silk screening. Chapter 5 discusses the importance of light, sound, and power control as communications, and Chapter 6 tells of power sources for the circuits in this text.

The circuits in the text were developed and tested with a proto-board. All circuits function exactly as described and provision has been made for further changes to enhance your exploration. All of the pinouts for the integrated circuits are from the *bottom view* as would appear if one held the IC in hand and viewed it from the bottom side. This was done throughout the book to make layout for etched circuits easier; however, the reversal must be in mind when making proto-board layout.

A sampling of circuits in this text have been constructed by students, both in proto-boards and finished etched boards; little difficulty was encountered.

Chapter 1

Introduction to Integrated Circuits

THE INTERFACING OF INTEGRATED CIRCUITS TO SOLVE ELECTRONics problems provides unlimited design possibilities. One needs to know (1) What ICs are available, (2) What are their characteristics and parameters, and (3) Are they compatible for interface utilization with a minimum of outboard components. The input and output drive characteristics of ICs are important to determine if direct drive is possible between integrated circuits or if a discrete component such as a transistor needs to be added to provide that balance. Reference manuals such as those listed at the end of the book are important to the designer to determine the characteristics and pinouts of the ICs to be incorporated in the circuit design. If these are not available, then one should secure information sheets upon purchase of the IC.

SCHEMATIC DRAWINGS

All the circuits shown in the book are drawn with the view from the bottom rather than the usual top view. Pin #1 is identified with a ′ symbol as a reminder. The reverse drawings were established for easier layout in the event one wishes to make a PC board directly from the schematic rather than develop a proto-board.

DESIGNING WITH PROTO-BOARD

The Design Mate I as shown in Fig. 1-1 was utilized in the development of all the circuits in this book. It is ideal because it has

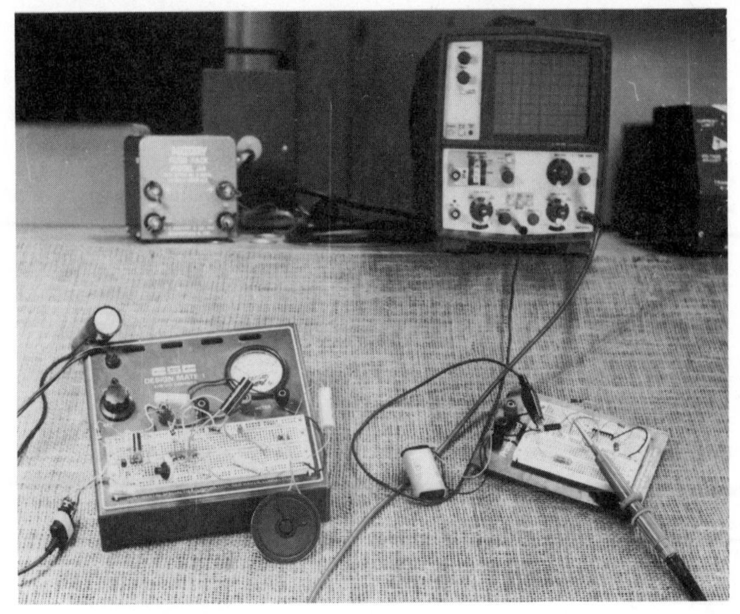
Fig. 1-1. Design Mate I and Proto-Board 102 in actual use with integrated circuits.

a regulated power supply from 3 to 15 volts, which was adequate for all the circuits. In a few instances there was a slight voltage drop because of circuit loading.

Proto-Board 102 works well for all circuits, however an external power source must be available. Some of the circuits called for a bipolar voltage supply; in these instances a second power source (battery or power supply) was necessary.

MEASUREMENT

In designing with ICs, one needs measuring instruments to determine interstage function, and/or troubleshooting. The digital multimeter is excellent for checking voltages, current, or resistances. There are times when both ac and dc values should be checked, therefore the digital multimeter must have both ac and dc capabilities for voltages and current. Such instruments as Beckman, Fluke, Simpson, or equivalent multimeters would be good choices. Beckman Model 310 was used with these circuits for analysis. The oscilloscope is important for determining frequency, time, and high impedance voltage measurements. It is exceptionally useful in determining time ON and OFF factors in the circuits.

CARE IN LAYOUT

Electronics is an area where the margin of error must be slight to insure proper function of a circuit. It is easy in proto-board layout to miss or make improper pin connections with jumper and component leads. The failure factor becomes greater as the number of ICs increase. Every IC has power-up pins for both positive and negative connections; it is wise to make these connections last otherwise the interconnections are made under power-up conditions and an improper connection of the other pins could destroy the IC.

It is wise in circuit design to layout by stages and then check, rather than construct a whole circuit and check it. In the former procedure, one must be sure the stage is functioning properly before going to the second stage. Likewise the second stage should be functioning properly; and both stages functioning properly before going to the third. This procedure is verifying the IC condition or testing the IC as well as the circuit. Remember, just because an IC is new doesn't necessarily mean that it is good. Another good check on first power up is to touch the IC with a finger; don't stay long as you can get a burn. If it is unduly hot there is a probability of an error in the circuit or a bad IC.

MONITORING STAGES

You will note throughout the book that LEDs (light-emitting diodes) are used frequently to monitor a stage function. This is a form of built-in tester and aids in the troubleshooting of the circuit in case of failure. LEDs can be designed into the circuit to consume little power yet serve as a valuable indicator that all is ok at that point. One can use various colors in the circuit for ease in recognition of a particular stage. Remember, LEDs are diodes and must be connected into the circuit with proper polarity.

COMPONENTS

Electrolytic capacitors are utilized frequently in these circuits for timing, dc isolation, by passing, coupling, and the like. They are susceptible to leakage in some amount and catastrophic destruction if connected with incorrect polarity. Remember the polarity is indicated by the symbol + and −. It is advisable to utilize tantalum type capacitors in timing circuits because they are a low-leakage type. Be sure all voltage ratings of capacitors are greater than the voltages at the point of insertion.

Switches are in circuits at a particular point to provide a unique function. In many instances they are omitted in the power supply input to save cost. A switch should be added in either the negative or positive lead of the power input to control the ON/OFF state.

READING THE CIRCUIT

The circuit explanation of function should be read very carefully along with an eye trace of the schematic to note what is happening from the beginning to end. Some users may prefer to finger trace the circuit along with the reading. Try to keep in mind what each stage is doing before going to the next.

FURTHER STUDY

There are many variables to these circuits that are affected by components. Changes can be made in components that will affect the circuit. It is a prudent idea to get the circuit to function first before experimenting by changing component values. However, learning is enhanced by being able to redesign a circuit through component changes. It would be wise to change only one component at a time and note the effect on the circuit. The circuit explanation will help to identify what the components are doing and may be self evident of changes that will affect the circuit function.

ICs AND DISCRETE DEVICES

This book has utilized CMOS ICs because they provide a wider range of voltages and are less susceptible to damage from overheat or misconnections. Occasionally, three point regulators are employed to provide different voltages in the same circuit. They are used often to insure accuracy of time when variance in supply voltage should take place, such as battery wear.

Discrete devices such as transistors, SCRs, or triacs are interfaced with ICs to handle unique requirements. Such devices are identified by number in the schematic; other types will work with adjustment of the bias resistors. Peripherals are identified or drawn as examples of how to extend the capabilities of the circuits.

INTEGRATED CIRCUITS

An integrated circuit can be a very simple array or a complex mixture of transistors, diodes, resistors, and capacitors to perform a particular function. A simple array would be two or more independent transistors that are not interconnected but are on a common subtrate. The complex integrated circuit is designed to perform a specific function like a voltage regulator, fluid sensor, audio amplifier, timer, and the like. It is possible to interface these integrated circuits in a variety of designs to extend the versatility even further.

Early integrated circuits were called SSIC to mean (small-scale integrated circuits). The art of manufacturing brought forth LSIC (large-scale integrated circuits) and has since advanced to VLSIC (very-large scale). The art now with ICs is approaching the super LSIC and the end has not been reached.

Both digital (ON and OFF) integrated circuits and linear types may be interfaced to provide unique timing, transducer activation, and power output. The option of utilizing discrete components such as power transistors, triacs, SCRs, and diodes in the interface with integrated circuits extends the range of design possibilities.

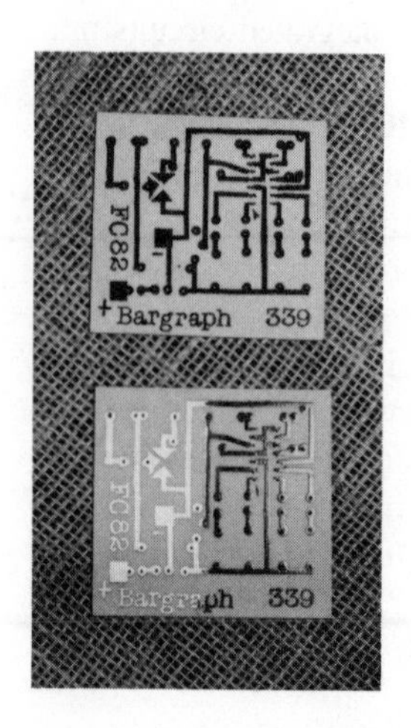

Chapter 2

The Etched Circuit Board

An etched circuit board can be made by hand drawing with resist pen, negative process, positive process, or silk screen. Regardless of the method employed, one needs to make a layout from the schematic to provide for the placement of components. The layout planning requires consideration for size of components, both off- and on-board placement, lead runs, and power input runs. It is advisable to make the layout resemble the schematic for ease in later tracing in the event of troubleshooting.

LAYOUT PLANNING

The first step in layout planning is to have the schematic close at hand for IC element identification. Since all ICs in the book are drawn in the schematic from the bottom view, the task is easier because reversal is not necessary.

The layout planning should take in consideration the final packaging because the copper clad board may need to be round, square, or some other geometrical shape. Problem #41, Safety-Flashing Light is utilized as the example for the layout-planning explanation. The flashing light will be in an amber pill box (found in any drug store) mounted on a plastic box. Approximately 7-feet of zip cord will be run from the box to a plug that will jack into a cigarette lighter. The plastic box is 4 × 3 × 1.5 inches and the etched board will be mounted inside the box. The above data tells you that a copper clad board 2.25 × 2.75 is the working size for the project.

LAYOUT OF THE BOARD

On a clean piece of paper or card stock 3 × 5 inches, draw the perimeter of the board on which the circuit will be etched. Look at Fig. 2-1 and Fig. 8-4 as you read the next ten steps.

1. Select an 8-pin pressure adhesive template and place on the card to represent the position of the LM555.
2. Draw the + and − bus runs at top and bottom.
3. Connect pin #1 to the negative bus.
4. Connect pin #8 to the positive bus.
5. Draw runs out from pins #7 and #6 to provide space for R1 and R2.
6. Draw a jumper from pin #6 to get back to pin #2.
7. Provide a spot for C1 to get from pin #2 to the minus bus.
8. Bring a run out from pin #3 with jumper to transistor Q1 base via R3.

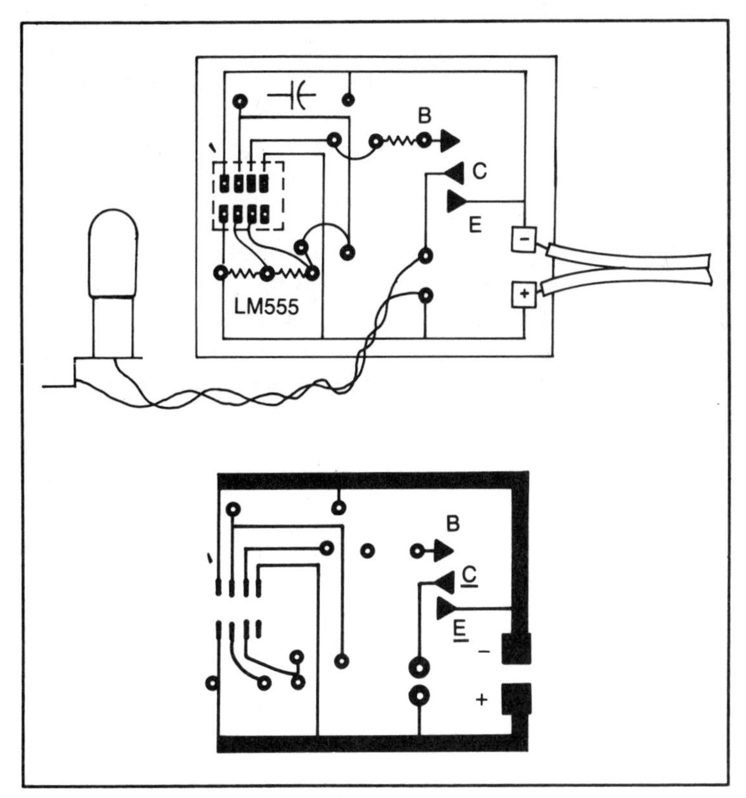

Fig. 2-1. Layout of project 41, Safety-Flashing Light.

9. Connect emitter to negative bus.

10. Connect collector of Q1 to lamp socket for lamp #47 and then to the positive bus.

Note that an attempt has been made to keep the layout to look exactly as the schematic.

TAPING THE ACETATE POSITIVE

The photo positive process is preferred because the taped acetate becomes a positive for direct exposure to a (positive sensitized) copper-clad board or to make a silk screen for a mass production run; no intermediate steps are necessary.

A clear sheet of acetate is placed over the layout and fastened in place with masking tape to hold registration while taping the circuit. Layout graphics are available in different tape widths, gauged pins for ICs, Ts, elbows, donuts, transistor, and the like to make the positive layout easier to construct. (See Fig. 2-2) Rub on letters and numbers are available to give the finished project a professional look.

One could draw the circuit directly on the copper board with a resist pen since the circuit is not too complicated. However, when several ICs are in the circuit or more than one board is to be etched, it is prudent to utilize the acetate positive method.

PREPARING THE COPPER CLAD BOARD

Presensitized positive boards are available for those who do not care to make-ready the boards; they give excellent results. A photo positive sensitized spray can is available for making the sensitized board; the process requires care and control to insure success. The following steps should be adhered to carefully.

1. Physically clean the copper board with 400 grit paper. You can follow this with a sink cleanser if you wish.
2. Wash and dry the board thoroughly.
3. Spray a coat of positive sensitizer on the surface with nozzle about 8-inches away. Make both horizontal and vertical passes. Two thin coats are ideal. It is important to cover the surface evenly. This should be done in a darkroom.
4. One can let it dry over night in the darkroom or force dry with an infrared lamp 12-inches away from the bottom side. Place it in a dark box upside down with infrared heat on the back side. It

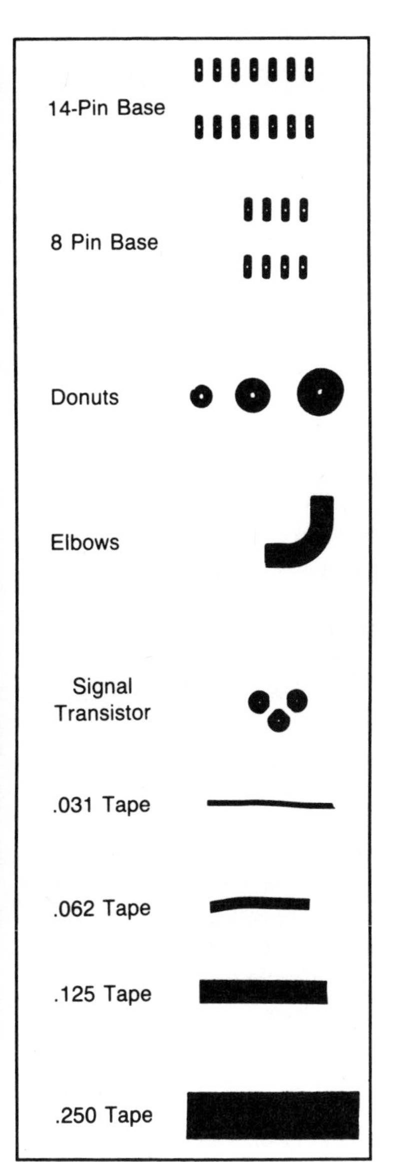

Fig. 2-2. Layout of graphic materials for taping acetate.

takes about ten minutes. Make a curved wire cradle so the board rests on the edges.

5. The sensitivity is not the same as photographic film so one can carry the sensitized board in incandescent lights for short periods. The board is most sensitive to ultraviolet light.

EXPOSURE OF THE BOARD

The acetate positive is placed on the sensitived copper surface and should be held in a frame to insure flush contact. The exposure can be made in sunlight or by artificial ultraviolet light. We used a sun lamp 8- to 12-inches from the surface and exposed it for about 4 minutes. Longer exposures will work but the tendency is to undercut. (See Fig. 2-3).

DEVELOPING THE BOARD

A commercial developer is available from electronic supply houses. It is a lye so handle with care and use in a glass or rubber tray as a developing container. Lay the exposed board in the tray and rock the tray gently until the outline appears; it takes about 2 minutes and you cannot overdevelop. The outline color of brownish-purple should appear. Wash thoroughly in cold water as this sets or hardens the outline. Dry with a paper towel and examine the surface. All outline edges should be sharp including letters or numbers. Hold at an angle to the light and the outline surface should be glossy if correct. You may wish to run some samples at different exposures to test the exposure time. Be sure to use glasses, rubber gloves, and tongs when working with the developer.

ETCHING THE BOARD

The board is etched in ferric chloride, which can be purchased in stock solution from a number of electronic suppliers. It can be etched in an open glass tray or one can utilize a spray etcher which brings the solution to about 110 degrees fahrenheit. It will etch in a cool solution but the etching time would be longer. Remember, that the part the light would not reach in the positive process is the resist and will remain. Ferric chloride is toxic so handle with care—again use glasses and gloves. Surgical gloves, which are purchasable in a drug store are excellent.

Wash the board in running water after the etching is complete. The next step is to remove the resist from the copper outline surface with fine grit paper. It must be clean before component soldering can take place or before plating. A tin plate solution is available for immersion plating (this is not electro-plating).

Plating with tin is recommended for a professional job. It inhibits corrosion of the copper and provides easier soldering because solder has an affinity for tin, which is an alloy element of solder.

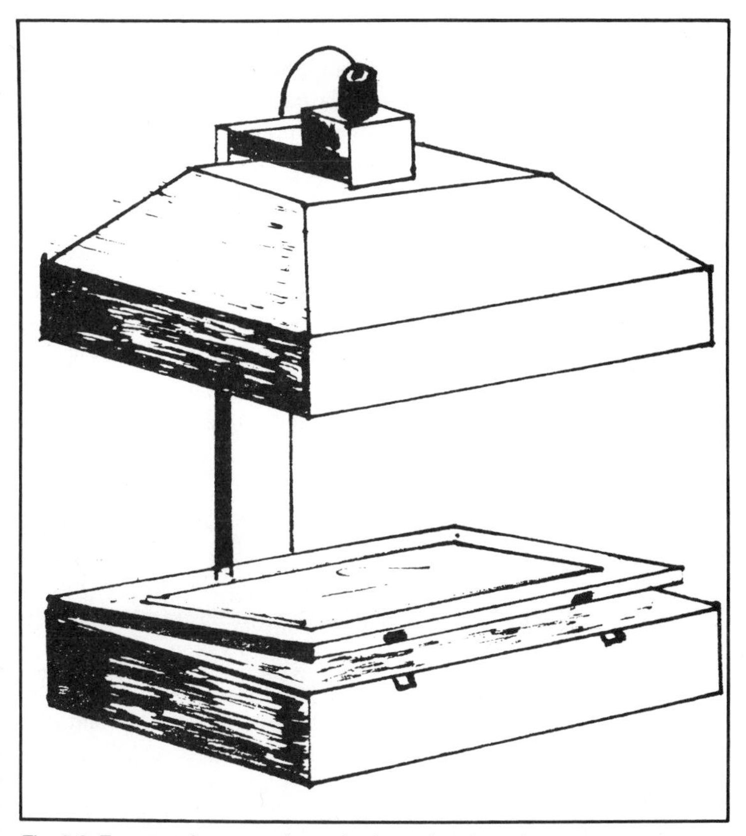

Fig. 2-3. Exposure frame, sunlamp shade, and sunlamp for exposing sensitized copper-clad board.

DRILLING THE BOARD

All the IC and donut spots on the board are drilled for mounting components. A high speed drill such as Dremel with a silicon carbide #60 drill is ideal for good clean holes. The drill breaks very easily so there can be no lateral movement of the board when drilling. The drilling action should be quick like a punch press rather than a slow down feed. Fiberglass board will cause short drill life with a standard high speed drill. Cobalt high speed drills are good but carbide types are better.

MOUNTING AND SOLDERING COMPONENTS

The major components are mounted first such as ICs, transistors, etc. It is advisable to utilize sockets when the major component has over three legs; this is particularly true of ICs. Misplacement of parts can be avoided by looking at the layout, particularly if the designer marks in the component symbols at the layout time.

The major components should be examined carefully for base identification to be sure they are inserted correctly; reference manuals or instruction sheets will be useful for leg number identification. ICs have a notch and an indent spot on top of the plastic to help identify pin #1. Remember, pin #1 from the bottom view was designated with a ′ mark on the schematic; be sure this mark is on the positive so it will be on the board. Figure 2-4 is an example of the indent identification and shows the relationship of the schematic numbers to the actual pin numbers. Inserting and removing ICs from sockets can be critical. Pins may be bent under and yet appear to be in the socket.

SOLDERING A PC BOARD

Soldering a PC board is a major skill. A solder pencil and 60/40 rosin core solder are two important factors. A heat-controlled solder pencil is preferred over a standard pencil, particularly when long soldering periods are required and pencil overheat would prevent good soldering.

The soldering of IC sockets requires skill and control with space tolerance that are very small. A steady hand and good hand and eye coordination are necessary. A magnifier head piece is recommended as an aid to give the designer the precise sight to prevent bridging when soldering.

A properly soldered lead would look "shiny" rather than gray; a gray joint indicates a cold joint or fracture from lead movement. The gray joint usually means a resistance point and is like inserting another resistor. Solder joints have been observed that were completely open circuits. This is a case where the joint is encapsulated with solder but there is no electrical connection.

Cleanliness is important in soldering. Some resistors or capacitors leads may be oxidized over time; one should scrape them clean in these instances or retin to brighten them. Be careful not to oversolder. I have observed oversoldering and overheating to cause the solder to run into the IC socket and thus close the entrance. If this happens, insertion will be impossible.

MAKING THE TRIAL RUN

The initial trial can be made by pressing the control switch, transducer change, or signal insertion. Project #26, Adjustable Tone

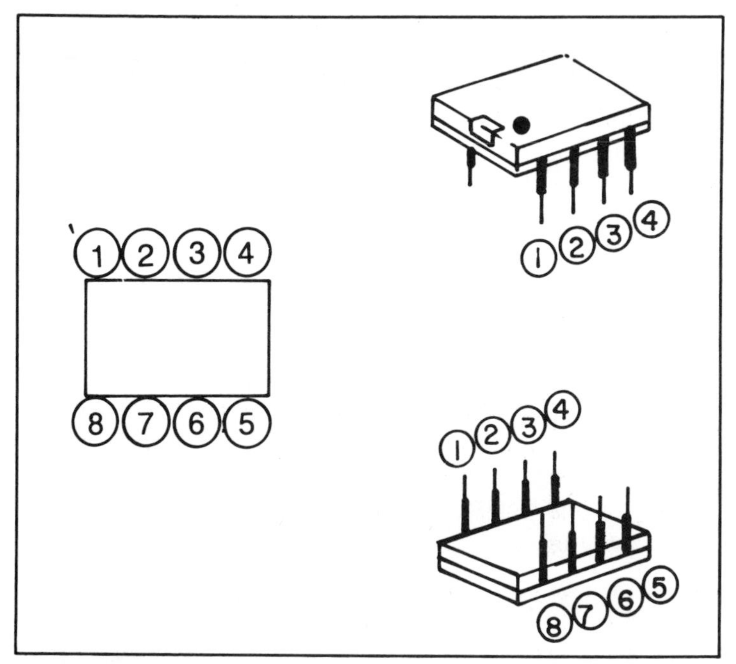

Fig. 2-4. Basing diagram in relation to physical integrated circuit for both top and bottom positions.

Generator-Amplifier, would make a signal generator testing device. The signal going to the speaker can be tapped as an output signal, however the volume control R3 should be set at the low end to prevent overdrive. This device will check all audio stages from output to input, or one can inject the audio signal at the input and trace it with an oscilloscope.

If a circuit doesn't function at all, the usual check is to voltage trace from each side back to the IC pins. Every IC has power-up pins, thus voltages should be measured from minus along the positive bus to the positive power-up pin, or from the positive along the negative bus to the negative power-up pin. Also go to the top side of the pin to check if current is passing through the socket. All other lead runs from the bus voltages should be checked to determine if the voltage reaches that component. Don't forget to check the voltage at the source under load.

Quite often a hairlike crack in the circuit board is unnoticed by the naked eye. This defect should surface with a voltage check. Flexing the board slightly while observing the copper runs under magnification will reveal the defects. Continuity checks of all copper runs with an ohmmeter to show infinite resistance will reveal open circuits. The hairline crack when found can be bridged with solder. Cracks or breaks that are greater than 1/32 of an inch should be closed with a wire jumper.

Resistors and capacitors cause very few problems if they are of the proper value. A color code misreading can be the one stopper to success; be sure to read the color code carefully. Incorrect capacitors will change the frequency or timing. A digital capacitor checker is useful as a check to be sure of correct value or bad conditions. Electrolytic capacitors are the usual suspect when a problem occurs. One must observe correct polarity to keep leakage to a minimum. Electrolytic capacitors of tantalum dielectric are preferred where leakage is a problem such as in frequency or timing circuits.

Finally, don't forget to substitute the major components. One can find a bad or marginal major component on occasions so keep these in mind as being suspect.

Chapter 3

Proto-Board Usage

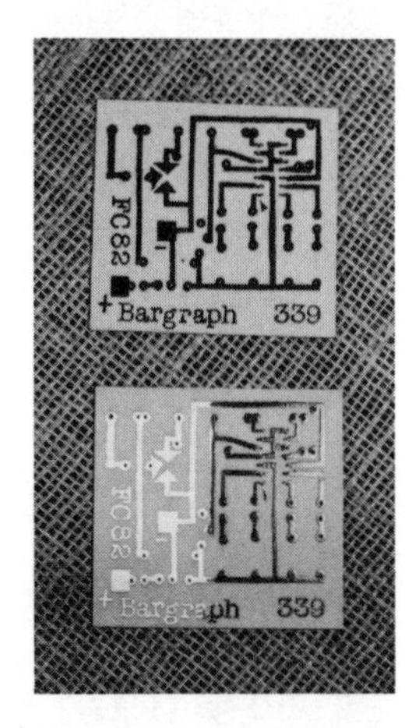

THE PROTO-BOARD IS A DEVICE FOR DESIGNING AND BUILDING circuits in the first phase of circuit development. It is easy to make changes without soldering or unsoldering because all connections are by socket contact. A variety of sizes are available; the more sophisticated types provide a built-in variable power supply with regulated output voltages.

THE PROTO-BOARD

Less expensive proto-boards like the Proto-Board no. 100, no. 102, or ACE 109 by AP are adequate for building the circuits shown in this book. The no. 100 board was utilized in an Institute for the Gifted and Talented by 4th, 5th, and 6th grade students. As many as four ICs were employed with 56 connections or more. The success rate was remarkable even though eye/finger coordination was not fully developed by these youngsters. Elementary teachers with interest in electronics but little background could wire circuits on these boards.

A close look at Fig. 3-1 shows the structure of proto-board no. 100. The center strip provides two voltage-buses for power input to the ICs. The upper and lower sections provide a series of 5-hole strips that are connected vertically with a space strip between them. This space strip provides the gauge width for insertion of ICs. The actual use of the proto-board will be explained later in the wiring of Project #41 (Safety-Flashing Light).

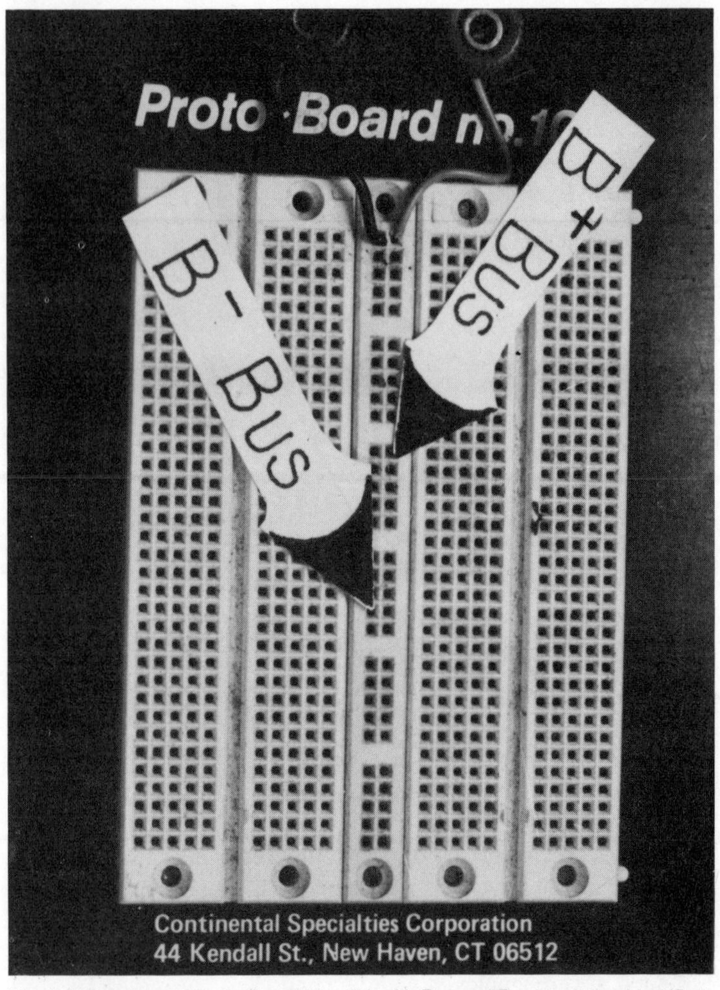

Fig. 3-1. Structure of Proto-Board No. 100 with B+ and B− bus strips identified.

A more expensive but convenient proto-board is the Design Mate I that is shown in Fig. 3-2. This board provides an adjustable-regulated output voltage that is metered. The bus voltages appear as a horizontal row at the top and bottom. One can use a jumper wire to provide a double row; or an auxiliary power supply can be added to the second row at top and bottom for a bipolar supply when one is required in the circuit. Remember the 5-hole vertical strips at the top and at the bottom are tied together so they are common. All strips are replaceable in the event of damage to a section.

WIRING THE PROTO-BOARD NO. 100

Wiring of a proto-board is straightforward if one starts from left to right and in the same symmetry as the schematic. Remember a proto-board is wired with all components seen from the top view.

Remember all the ICs in the schematics in this book were drawn with the bottom view to make the etched circuit layout easier.

The designer will need to think reversal when wiring the proto-board; the top pin numbers on the schematic will be toward the bottom of the board and the bottom pin numbers on the schematic will be on the topside of the board. This is not too difficult: just wire from the respective pin numbers to the runs.

One will need to cut and strip 1/4-inch from the ends of #22 wire as jumper leads. It is a good idea to precut some 3/4-inch, 1-inch, 1-1/2-inch, 2-inch, and 3-inch wire pieces to have ready for wiring. Some designers prefer red, black, yellow, orange, and green colors for better identification and later recheck visibility.

Look at Fig. 3-3 and the schematic to trace the following steps:

1. Orient the proto-board so the power input is to the left.
2. Take the LM555 and place it on the board by bridging the

Fig. 3-2. Design Mate I Proto-Board showing meter, voltage-adjust control, and B+/B− bus strips.

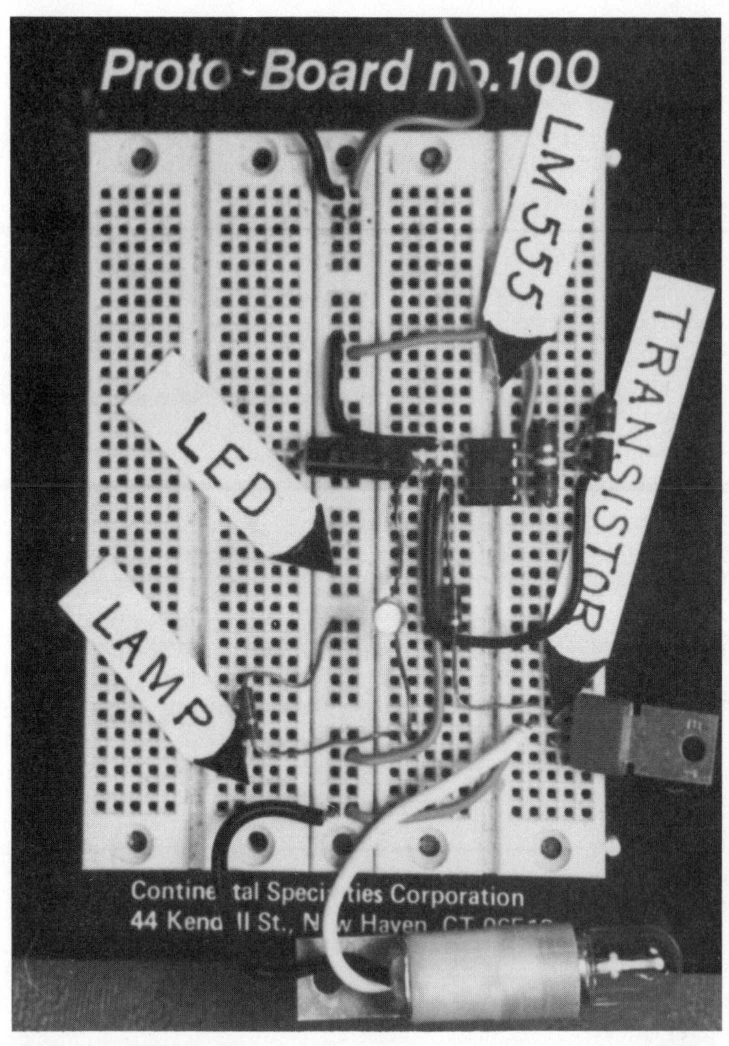

Fig. 3-3. Proto-Board No. 100 showing project 41 development with selected components identified.

white space gap. You may have to "wiggle" it to seat firmly.

3. Take a red jumper wire and connect from pin 8, B+ input to B+ strip.

4. Take a short black jumper wire and connect from pin 1, B− input to B− strip. Be meticulous in dressing leads; keep them flat on the board.

5. Take a jumper wire and connect from pin 2 to pin 6. Note that it is dressed around the IC rather than over the top; you may

need to replace the IC later so don't lock it in with over the top jumpers.

6. Take a 2 μF.,C2, capacitor and connect the + end to pin 2 and the other end to the B− bus.

7. Take a red jumper wire and connect pin 4 (enable pin) to the B+ bus.

8. Take two 100 k resistors, R1, R2, and bend to form a loop. Connect R1 between pins 8 and 7; the R2 between pins 7 and 6.

9. Take an LED and note the flat side, which is the cathode end. The flat side goes to an unused 5-hole vertical strip and the other end to pin 3 of the IC. Take a 1 k, R4, resistor and connect between flat side of LED lead end and the B− bus.

At this point the IC can be checked for function. Apply a voltage between 9 to 12 volts to the proto-board and the LED will flash. This is the monitor part of the circuit; every time pin 3 goes high the LED will turn ON.

10. Place the pnp power transistor, Q1, 2SC1061 into 3 holes (side by side) as shown in Fig. 3-3.

11. Take a 10 k resistor, R3, and connect between IC output, pin 3, and the base pin of Q1.

12. Take a red jumper lead and connect between emitter and B+ bus.

13. Take a #47 lamp in socket and connect between collector and B− bus.

14. Bring up power slowly to 12 volts as designed. When the LED is ON, the output pin 3 is high; thus Q1 is reversed biased and the #47 lamp is out. When the output at pin 3 goes low, the LED is out and the lamp is ON because Q1 is forward biased.

Lets assume you have a good unmarked pnp power transistor that you would like to use rather then purchase the one specified. This can be done experimentally. Insert it into the circuit and bring the power supply voltage slowly toward the 12 volts. If it doesn't flash at all, it could be bad or its beta or gain is too low. If it flashes dimly, then it is ok but the gain is too low. Change the bias resistor R3 (10 k) to 4.7 k and observe the lamp; if the brightness of the flash increases but not to full brightness, the bias resistor is still too large. Continue to reduce R3 in value until full brightness appears as it flashes.

Another way to design around the unknown pnp transistor is to check it out separately. Disconnect R3 from pin 3 of the LM555. Place a dc voltmeter across the lamp. Connect the open end of R3 to the B− bus and note the voltage drop across the lamp. Now decrease the value of R3 in steps and note the rise in voltage and increase in brightness. When the voltage reaches 6.3 volts across the lamp, the bias resistor is about the correct value. If the input voltage from the supply is 12 volts, then 6.3 across the lamp and 5.7 across the transistor collector to emitter should be observed. The new R3 becomes the bias resistor between pin 3 of the IC and the base of Q1.

Can you use an npn power transistor instead of a pnp? Yes, but you must change the circuit to connect the lamp end to the B+ and the emitter to the B−. The light will flash at the same time as the LED. When the output of the IC goes high, the LED turns ON the transistor is forward biased and the lamp is ON.

WIRING THE DESIGN MATE I

The Design Mate I or similar boards like AP Model 101 will be wired in the same manner. They have the convenience of a built-in adjustable-regulated power supply. One advantage of an adjustable supply is the warning affect of impending troubles. When overload occurs from misconnection or component failure, the power supply will not come up to voltage.

The wiring of the circuit is the same as explained for proto-board no. 100. The singular difference is the placement of the B+ bus at

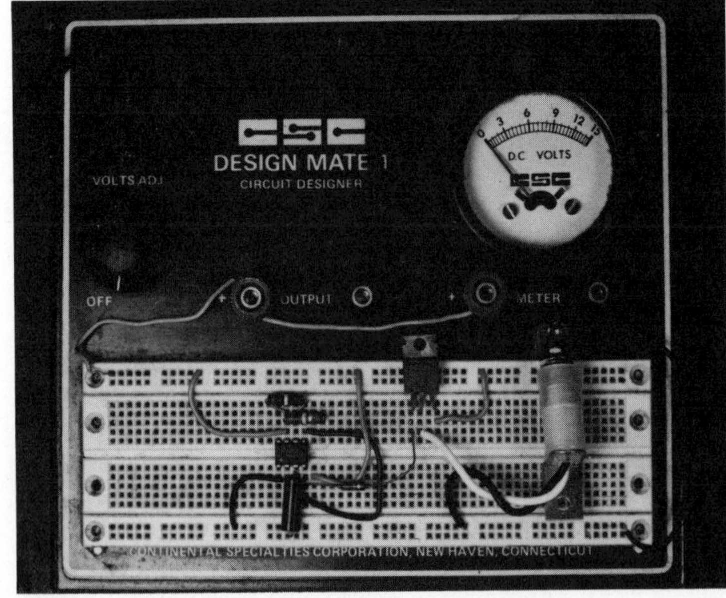

Fig. 3-4. Design Mate I showing completed circuit development of project 41.

the top and the B− at the bottom rather than in the middle. There is less congestion in this particular circuit as shown in Fig. 3-4. Note the indent or spot to identify pin 1 of LM555.

Potentiometers, switches, humistors, thermistors, LDRs, and the like can have leads soldered to the pins for outward extension where necessary and to provide connection to the proto-board.

Chapter 4

Silk Screening Circuits

THE PHOTO POSITIVE PROCESS TO MAKE ONE OR TWO BOARDS IS an excellent method. However, when several boards are to be constructed or multiple boards as in quantity production, then silk screening is a good procedure. Excellent results with fine details can be achieved with silk screening.

SILK SCREENING METHOD

Silk screening of the copper clad board to deposit the circuit resist is a stencil type process. Resist ink is squeezed through a form to deposit the circuit outline on the copper surface. The outline with resist covering remains when the board is etched with ferric chloride. Figure 4-1 shows several circuit boards with resist in place after etching.

After etching the board, the resist is removed and the copper circuit outline is drilled and plated prior to mounting the components. Figure 4-2 shows a board with part of the resist removed to expose the copper outline. Note the board has been drilled; the drilling can be done prior to removal of resists.

The resist can be removed chemically with lacquer thinner or physically with fine grit paper. The surface must be cleaned to insure good plating or soldering of components if plating is not included. Figure 4-3 is a plated board that is ready for mounting of parts and soldering.

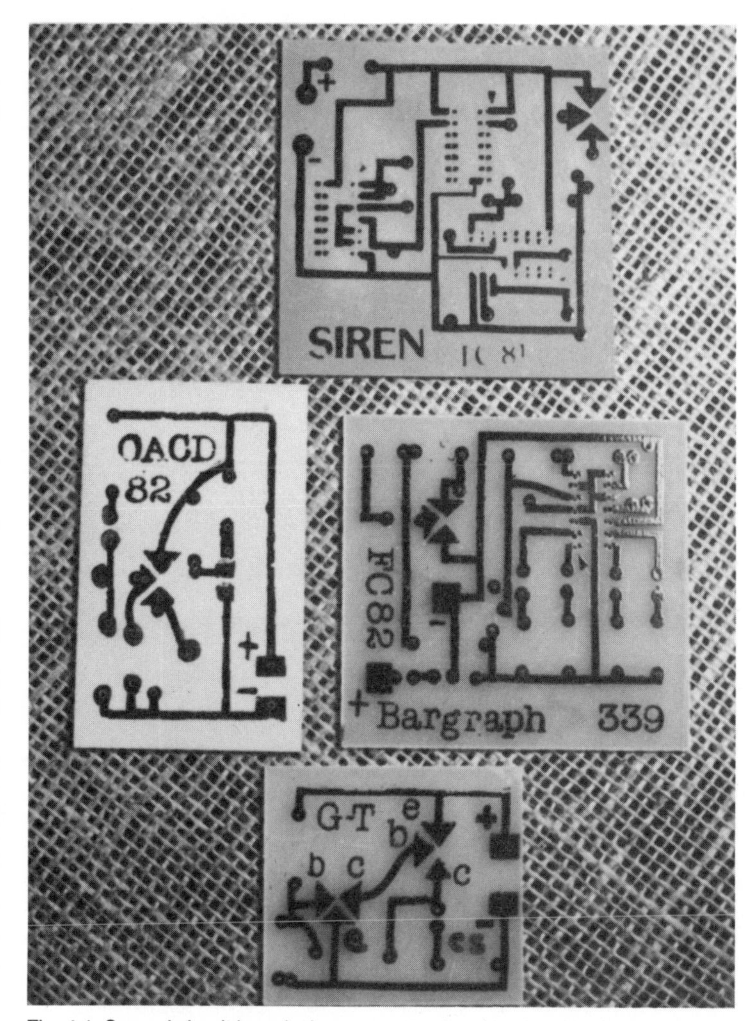

Fig. 4-1. Several circuit boards that were completed by the silk-screen process.

SILK SCREEN EQUIPMENT

The silk screen equipment consists of a backing board, a rectangular frame with silk screen attached, hinged clamps, and a side stop to keep the screen away from the printing surface when not in use. The screen comes in various measures of fineness of mesh like 10x and 12x. It can be used many times if properly cleaned after each use. Figure 4-4 shows the screen frame with registration template on the backing board.

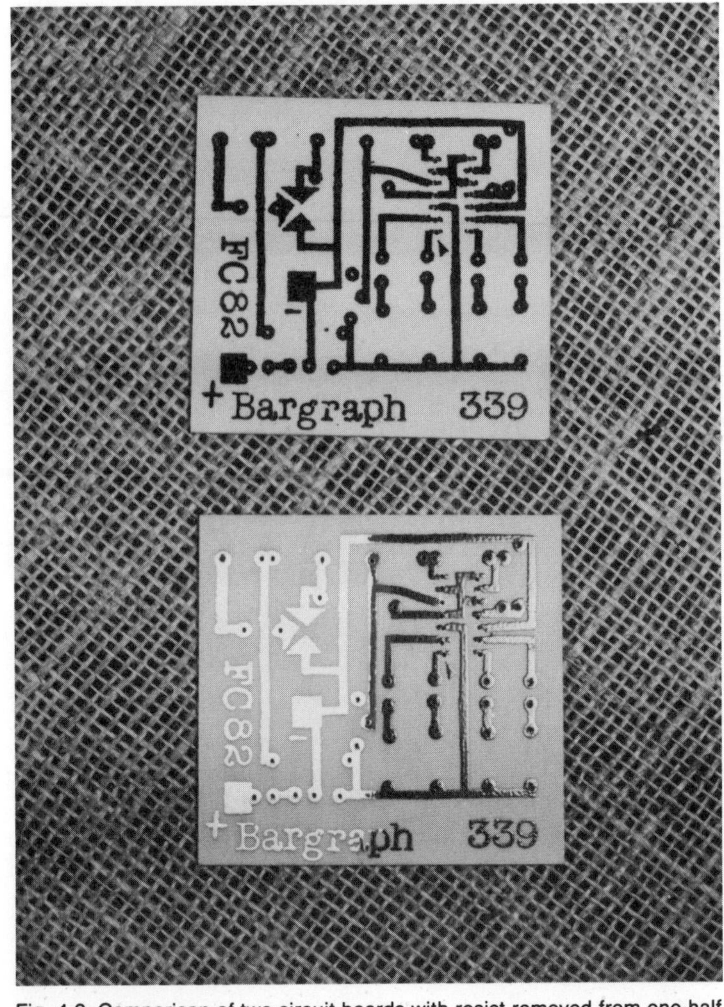

Fig. 4-2. Comparison of two circuit boards with resist removed from one-half of the board after etching.

PREPARING THE SCREEN STENCIL

The circuit to be duplicated can be prepared as a screen stencil by hand cut or photo screen process. In the hand cut process, a sheet of stencil material is taped over the layout and the circuit outline edges are cut with a fine knife. The gelatin is lifted from the surface to show a light area that is the acetate backing. The top picture in Fig. 4-5 shows a hand stencil. The hand cut stencil is appropriate when the circuit is not complex. One needs a light box for the layout and the stencil to be viewed properly for cutting.

The photo process requires photo sensitive-stencil material that is available from a printing supply house. It comes in both water, and hydrocarbon solution solubility. It is often referred to as Poly Blue. You will need the water soluble type since the resist ink is of hydrocarbon composition. It must be stored in a dark place because

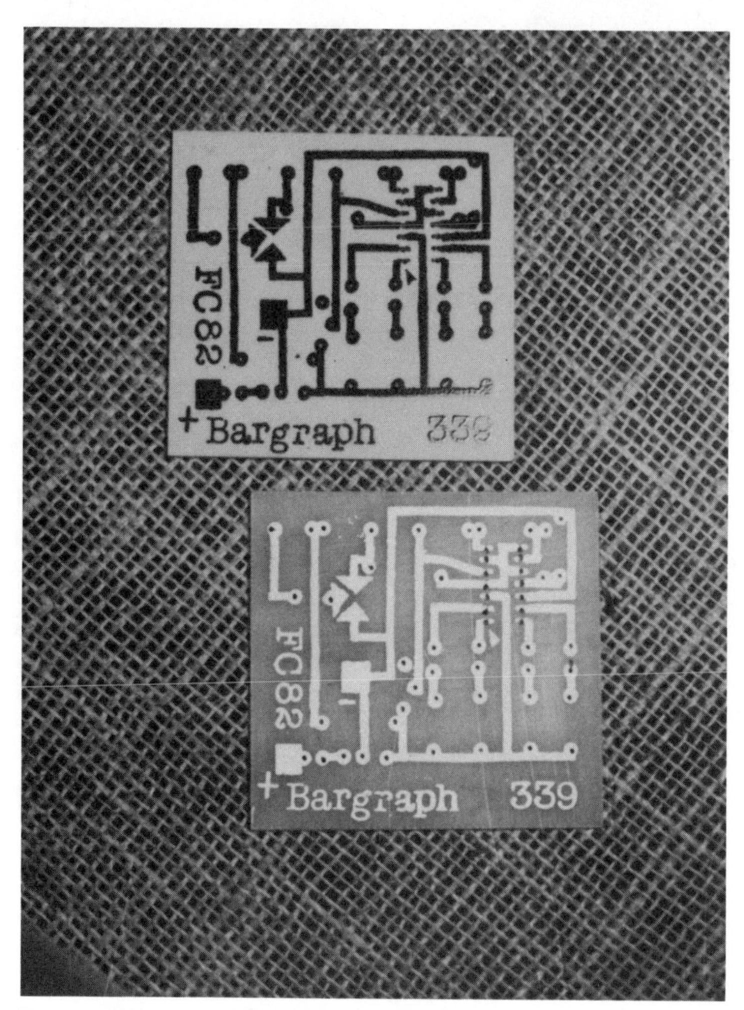

Fig. 4-3. Comparison of two circuit boards with resist removed from one board and tin plating completed. Holes have been drilled for mounting of components and soldering.

Fig. 4-4. Silk-screening frame with registration template on the backing board.

it is light sensitive. If you examine the stencil material, you will note it is a gelatin-type coat on acetate backing. The emulsion side is dull whereas the backing is shiny. The emulsion coated side will adhere to the screen in making the silk screen for printing.

The photo stencil is prepared by placing a sheet of the material in the light frame with the emulsion side down. Place the acetate positive (as made and explained in Chapter 2) on top of the photo material with the taped side down. Expose the two layers with a sunlamp for two minutes. The photo material is more ultraviolet sensitive than incandescent sensitive, therefore you can work with it without a darkroom (subdued light is recommended). Remove the photo material after exposure and develop. The developer is a hydrogen peroxide solution of 1 part hydrogen peroxide to 8 parts of water. A commercial developer is available. It will take about 1.5 minutes in the developer with agitation. Keep the emulsion side up in the developer to prevent scratching; also, you can observe a peel back at the edges of a grayish reaction. When the peel back is about 1/8 of an inch, the development is complete. The next step is to wash in running tap water of about 68 degrees (temperature not critical). The part of the light could not strike will wash away to expose the acetate backing. Figure 4-6 shows the completed stencil with the emul-

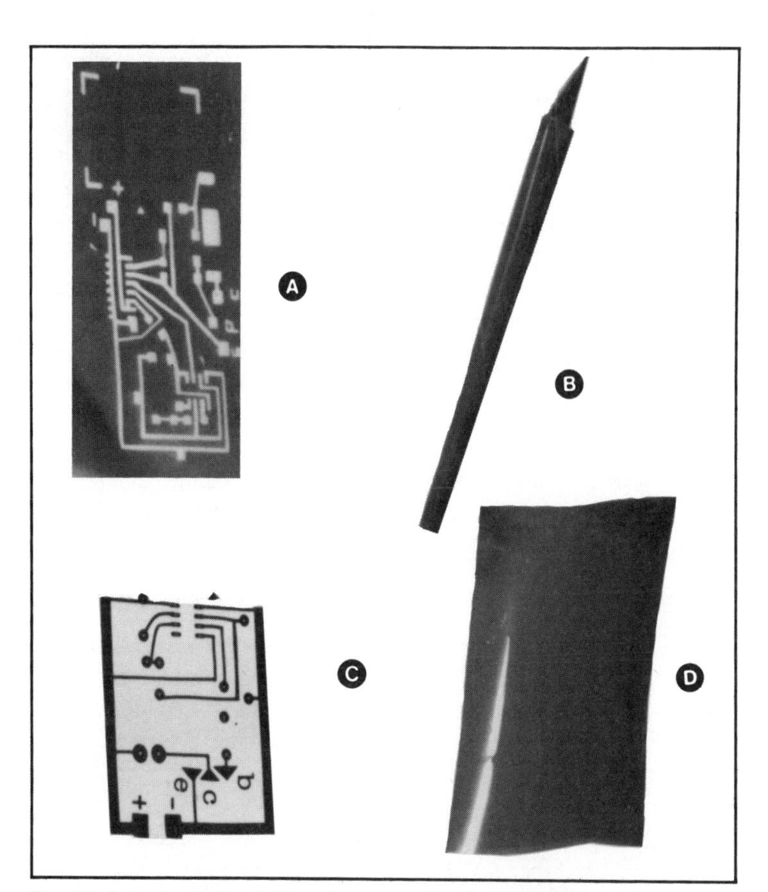

Fig. 4-5. A hand cut stencil (A) and proper cutting knife (B). The positive acetate layout (C) to be used with photo-sensitive film (D).

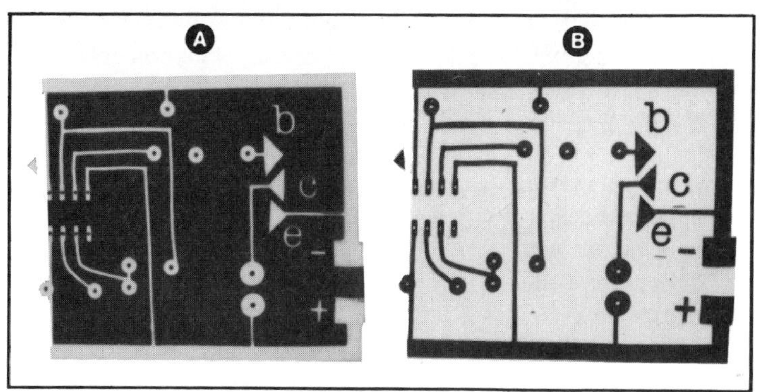

Fig. 4-6. Completed photo stencil (B) with the emulsion side up. Taped positive layout (A) shows comparison. Note: where the light was not able to strike the sensitized emulsion, the emulsion is removed when washed.

sion side up. Note it is the same as the positive layout next to it.

Ulano Corporation has CDF2 film that is recommended for PC board work. This process is different by way of procedure; no developer is necessary with this film. The film is adhered to the wet screen prior to exposure. The emulsion side is placed on the screen and a squeege drawn across the other side to force the emulsion into the screen. After the emulsion dries, the support sheet is removed. The positive is placed with taped side down on the side where the support sheet was removed and exposed with ultraviolet light. About 4 minutes was the exposure with ultraviolet light about 8-inches from the surface. One might try test strips to determine exposure time for a particular light source. No developing chemical process is necessary or water controlled temperature during wash out. After exposure, the film is washed immediately until the circuit opening is clear. The screen preparation prior to exposure must be done in a darkroom.

ADHERING THE PHOTO STENCIL TO THE SILK SCREEN

The photo stencil should be adhered to the silk screen while the emulsion is still soft from the washing. Begin by wetting the silk screen with water. Place the silk screen frame on a flat surface with backside toward you. Place the photo stencil on the backside of the screen with the emulsion side toward the screen. Do not make any lateral movement in this step.

Place paper towels on a flat surface and turn the screen over to place it on the paper towels. You will see the outline through the screen. The next step is to get the emulsion to adhere to the screen. Place a paper towel over the outline and press lightly. Evenly press the emulsion to the screen. Hold the light and look through the emulsion. If it is correct there will be even contact with no bubbles. Be careful not to distort the photo stencil with too much pressure.

Set the completed screen aside to air dry until the emulsion has solidified again. You can tell by lifting at the edges. The acetate should lift to leave the emulsion on the screen. It is a good idea to let it stand overnight. Peel the acetate from the photo stencil very slowly. See Fig. 4-7 for an example of the adhered photo stencil.

MASKING THE SILK SCREEN

The silk screen must be masked to direct the ink only through the photo stencil when printing. The silk screen can be masked with cardboard as shown in Fig. 4-8, or you can purchase a liquid for

Fig. 4-7. Photo stencil adhered to the silk screen. Note: this is a used screen because traces of previous circuits are visible.

brushing in the mask. A hole is cut in the cardboard to be over the photo stencil and then taped with masking tape to prevent leakage. The cardboard is taped to the frame also.

REGISTRATION OF THE COPPER CLAD BOARD

A resist stencil ink is available from electronic suppliers. You piece of cardboard has an opening cut from it to receive the copper clad board. Such a template is shown in Fig. 4-9 with corresponding

Fig. 4-8. Silk screen masked with cardboard ready for printing. Note: the hinged clamps makes removal of the frame easy.

finished etched board to demonstrate size. Note the hinged clamps for adjusting the screen frame. This equipment design has the additional advantage of easy frame removal for cleaning. You need to position the registration template with silk screen in place to insure proper registration. The template can be fastened to the backing board with masking tape.

PRINTING THE COPPER CLAD BOARD

A resist stencil ink is available from electronic suppliers. You will need a putty knife, mixing pad, resist ink, paper towels, and solvent as shown in Fig. 4-10. The consistency of the ink is critical; if too thick it will clog the screen quickly in a few runs; if too thin it will run and cause bridging. The consistency of medium heavy cream is about right. Premix it on a pad and then apply to the head area of the screen; a little goes a long way.

The printing is straight forward. All the copper clad boards should be cleaned and stacked prior to inking. Place a blank in the registration opening; lower the screen to the board, and pull the ink across the opening with a squeege. Only one pass per board. You will note the screen is up from the board about 1/16 of an inch. You are pressing the stencil to the board as you draw across the opening. This means the screen will lift as you pass the board. The screen is being pressed to the board with the squeege as the ink is deposited. The lift action will prevent the board from sticking to the screen. Remove the printed board and check for completeness. Any incomplete boards can be recleaned and reprinted.

DRYING THE RESIST

An infrared heating lamp will hasten the drying of the resist. Place

Fig. 4-9. Registration template on backing board. The cut opening must be the same size as the copper-clad blank.

Fig. 4-10. Materials necessary for preparing the ink for printing (mixing pad, putty knife, resist ink, paper towels, and lacquer thinner or mineral spirits).

the printed boards under the lamp for about ten minutes to insure complete drying. Minor discrepancies can be corrected at this time such as filled in donuts or too wide a line. Minor defects can be corrected with a resist pen or unwanted fills can be scraped away with a knife.

CLEANING THE SCREEN

Cleaning the screen after a production run is messy but necessary. Dried resist will mean a nonreusable screen. Place the screen on several paper towels and pour lacquer thinner over the opening from the topside. This will clear any residue ink from the screen. Remove the cardboard mask from the screen. This can be replaced if one decides to use the screen again. Touch-up cleaning after mask removal will be necessary. Mineral spirits may be used (for safety) in place of lacquer thinner.

REMOVING THE STENCIL FROM THE SCREEN

If you do not plan to use the stencil again or you will make a new one when appropriate, then remove the stencil from the screen.

Since the stencil is water soluble; place the screen under a faucet of warm water and scrub both sides of the screen with a sponge. The stencil will soften and wash away to make the screen available for future use.

Chapter 5

Light and Sound as Communications

THE WORD COMMUNICATIONS COMES FROM THE LATIN WORD "communis" which means common. Communications is the process of establishing commonness with someone in transmitting a message that has meaning to the receiver and may solicit some response. The higher level of commonness between the sender and the receiver the greater the chance of completing the communications. Educational background, employment, travel, interest, abilities, and experiences are shared experiences of a common culture. The higher the level of intersubjectivity between the sender and the receiver in this commonness, the greater the chances of completing the communications.

Light and sound are elements of communications that transmit messages in different forms of meaning. Humans perceive light and sound transmissions and assign meaning to them. There are many different kinds of light and sound transmissions that have been defined and described to become knowledge of a common culture and solicit a common behavior.

Humans perceive messages by sight, sound, touch, taste, and smell. These are important in the process of developing a level of understanding of communications. We interpret data and assign meaning and value through definition, description, differentiation, and discrimination of the input. Sight and sound are the primary means for humans to receive and interpret data.

LIGHT AS A MESSAGE

Light plays an important role in communicating to humans. A light in its simplistic form is associated with the state of ON or OFF as indicated by dial lights in radios, televisions, recorders, and the like. The light that marks the doorbell button is an example of a location function. There is the key location light in the automobile or lighted letters on the shift location to identify engagement in proper drive. What about the highbeam light or oil pressure light on the instrument panel. Don't forget the stereo indicator light in the radio to let you know that stereo transmission is being received. There is a blinking light to remind you to fasten your seat belt. There is the light on the telephone in the office that indicates in multiphone connections that a line is in use, or the blinking light to signify that a call is on hold. Don't forget the ready light on a floppy disk drive or the blinking error light, or the printer interface light that turns from red to green to indicate readiness. In thousands of ways, lights communicate to persons each day and it requires knowledge of what they mean.

Light communication can be from person to person, person to device (machine), machine to machine, machine to person, and other interfaces. The traffic light is a good example of machine to person. Stop lights on an automobile, and blinking direction lights give warning to others through light communications of person to machine to person. Ambulances, police, and fire vehicles with flashing lights communicate the need for right of way. Monitors in hospitals and industry are ways of communicating to humans the nature and condition of an event. Navigations lights for channel markers and obstructions, boat mooring or underway lighting are necessary means of communicating to provide safety. How would airplanes land and depart from a busy airport without all types of directional lights. There are countless ways in which lights transmit meaning for convenience and safety.

SOUND AS A MESSAGE

The human organism builds a memory of sounds over time that distinguishes noise from meaningful sound that has a particular message. Communications by light requires more directivity if the message is not to be missed. Although the ear has directivity the cone of reception has a wider latitude. You become conscious of certain sounds when trained to do so. The differences in sounds are such that it is easy to distinguish the meaning of particular sound. A siren, automobile horn, truck air horn, tug boat blast, or train whistle are common and easily distinguishable. Not only do we hear tone differences but also repetition rate and intervening nulls.

In some instances the sound is merely used to get attention rather then having a meaning, or sometimes it is both. The seat belt fastening buzzer or the key in the ignition are reminders as well as providing a message. The rear door bell can be distinguished from the front door bell. The smoke alarm is different from the entrance monitor. The phone in one office can have a different ring than the phone in an adjacent office.

One can be trained to distinguish different sounds and to sort out the differences and assign meanings. It is not unusual for a person to hear several morse code signals, yet follow one because it has a slightly different rate and rhythm.

LIGHT AND SOUND PATTERNS

Light can have a pattern in transmission like the ON time is longer then the OFF time. It is possible to have a rhythm such as three short light bursts, three long bursts, and three short bursts then delay off to give the light message of S.O.S. Light can be flashed in different colors to impart a different message by color alone. The traffic light of *go, caution*, and *stop* is an example of color discrimination to impart meaning. Lights can be flashed in count to express meaning like three flashes, delay off, and then repeat. Light can change in intensity like half bright to full brightness and then back to half brightness. Lights can be flashed to indicate movement or direction like left to right or top to bottom. Lights can be sequenced to form letters, numbers, or other symbols. They can be additive or subtractive to give change in pattern.

Sound has greater latitude of tone and rhythm. The tones can be multiple sequenced to produce chord patterns. Alternate tone dispersion can be accomplished to give emphasis. The dispersion angle can be changed to give directivity. Sound can be transmitted in the same manner as light with repetition and rhythms. Loudness or softness can be interchanged to produce an effect on meaning. Sounds can be verbal and carry a message in a common language, scrambled and descrambled for message control, or coded to communicate directly with a particular receiver who has the key for decoding.

POWER CONTROL AS COMMUNICATIONS

A person or device communicates to other devices (machines)

in power control. In some instances the command is given by a sensor to an integrated circuit, which in turn communicates to a power switch. Automatic control is a process of a device sensing a change in conditions or situation and then controlling another device to perform a function. Feedback can be in the process to reorder the change. Feedback is the process in communications where the receiver sends information back to the sender to correct or refine a situation. A good example of feedback is a heat sensor that communicates to a device to turn on a fan that circulates air back to the sensor, which in turn causes shut down. There are examples in this book of integrated circuits communicating to other integrated circuits to control power.

BIDIRECTIONAL AND UNIDIRECTIONAL COMMUNICATIONS

Bidirectional communication provides alternate roles of sender and receiver communicating with each other. An intercom circuit in which either party can initiate a contact and carry on alternate receiver/sender roles is a good example of this procedure. Telephone, CB radio, marine radio, airflight radio, trucking dispatch, police, and ambulance transmissions are bidirectional communications. Opportunity for immediate feedback is possible in this form of communications. Flag and light signaling can be in this form of bidirectional communications.

Unidirectional communications means one way message transmission. Radio and Television are good examples of the unidirectional process. This type of communications means the sender has little contact with the receiver. Public address announcements are examples of the unidirectional method. The message must be all inclusive of the informational, motivational, and instructional elements. Feedback is latent in form and has no immediate reordering of the message. This book has many circuits of unidirectional communications with light, sound, and power control.

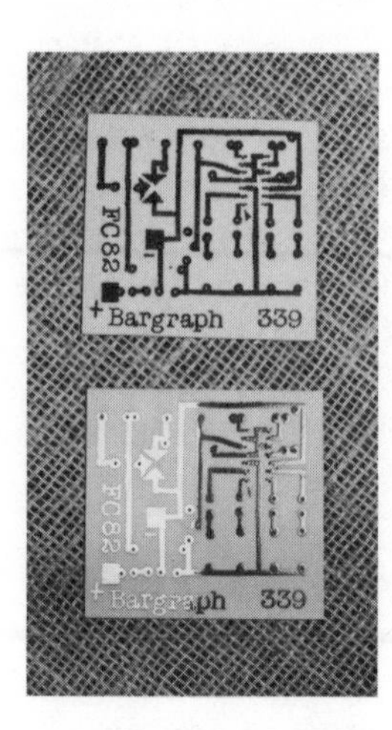

Chapter 6

Power Sources for Integrated Circuits

THE INTEGRATED CIRCUITS USED IN THIS TEXT CAN FUNCTION with battery power. In fact, battery power eliminates the concern for electrical shock. Rechargeable ni-cad batteries reduce the cost of battery replacement by reuse many times after recharging. Convenient wall receptacle chargers make the use of ni-cad batteries a good choice. Both proto-boards and completed etched circuits need energy to function.

PROTO-BOARD POWER

Proto-boards with built-in variable power supplies are ideal for circuit designing. They are available in both large circuit capacity and large power capabilities. The on-board power converters are adjustable-regulated types that produce stable dc for varying current loading. It is easy to construct an outboard power source for those proto-boards that require external power source.

POWER SUPPLY NEEDS

The circuits in this text require variable dc voltages although many circuits operate from a 9-volt fixed source. In addition, some circuits require a bipolar supply rather than single ended, which means a + and − dc supply with a common ground. The power capability must be variable also to meet different requirements. The better solution is to construct an ac power converter that is over-built in capability and then run it at an underrated level. One can never increase the dc level above the basic design but it is possible to reduce voltages by fixed and variable regulators.

POWER SUPPLY DESIGN CONSIDERATIONS

The primary consideration in designing an ac converter is the transformer. Transformers are rated as to rms voltage at the secondary and by current capacity. A transformer of 12-volts rms and 1-ampere capacity would meet the requirements for most circuits. Remember the dc output level follows the PV rather then the rms, therefore a 12-volt transformer secondary would have a dc capability at full-wave conversion of about 16.96 vols. (12×1.414). The diode losses in a full-wave design would be about 1.2 volts, and that would result in a supply of about 15 volts. An adjustable supply from 3 to 15 volts would be excellent for the ICs listed.

Diodes for the conversion are rated to PIV, PRV, PVB, DROM, which represent the backside (anode) nonconducting potential that cannot be exceeded. A 12-volt rms secondary in a bridge conversion would have a PIV of about 17 volts (12×1.414). Diodes like the 1N4002 and 1N4004 have a PIV that would exceed this characteristic of 17 volts. Similarly, one needs an I_f (forward current) rating of at least one ampere.

The filter capacitors are essential to convert the full-wave frequency to a steady dc voltage with low ripple. The filter capacitor must have large capacitance such as 2500 μF or better and a dc working voltage that is greater then the load dc voltage. In this case, 16 volts or greater would suffice. Polarity consideration is a must to prevent deforming of the dielectric and destruction of the component. Remember the curved side of the symbol is negative. The exact voltage can be achieved by utilizing three point regulators. They are integrated circuits with current limiting and thermal protection that are easy to use. In addition, they have the characteristic of ripple rejection, which is most important in audio circuits. One could design a circuit with plug-in capability to reduce the voltage to an exact quantity. The 78XX are positive regulators whereas the 79XX are negative types. The 78XX series will be utilized for the most part because most of the power requirements are positive source. The 7805, 7808, and 7012 regulators should be available to meet particular needs. The 78LXX series may be utilized when the current requirements are under 100 mA. Circuits to produce regulated variable supply are shown in this chapter, which is the more convenient system. All regulators should have a .3 μF disc capacitor at the input if the regulator is farther than 2 inches from the filter capacitor. They are necessary to bypass transient voltages that would otherwise damage the regulator.

SINGLE ENDED VARIABLE CONVERTER

A single ended variable supply with regulated output is shown in Fig. 6-1. This supply will produce from 4 to 16 volts with maximum current of 1.5 amperes which is more than adequate for the integrated circuits.

BIPOLAR CONVERTER—REGULATED OUTPUT

There are several integrated circuits that call for a bipolar source. The one shown in Fig. 6-2 would provide for this need. Note that both a positive and negative type regulator is needed.

SINGLE ENDED CONVERTER

A single ended converter with selectable voltage options is useful as shown in Fig. 6-3. A switching technique provides a choice of fixed voltages with regulation.

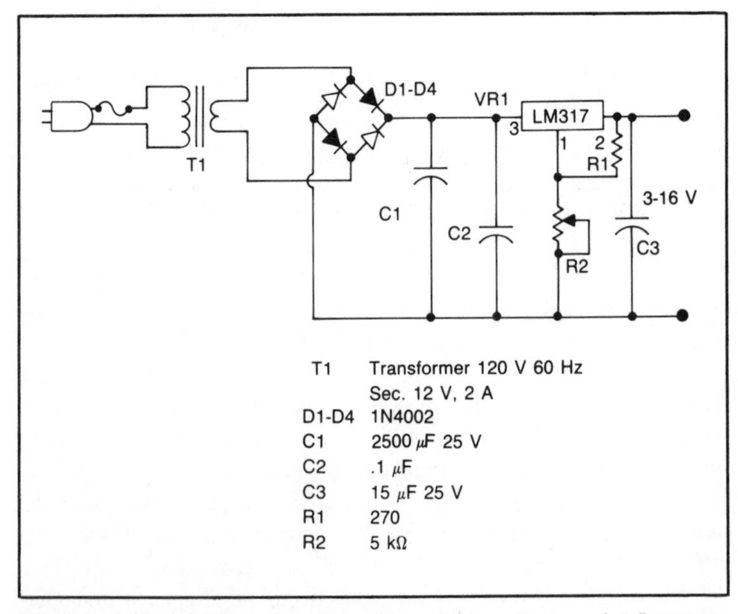

Fig. 6-1. A single-ended converter with variable voltage and 1.5-ampere capability.

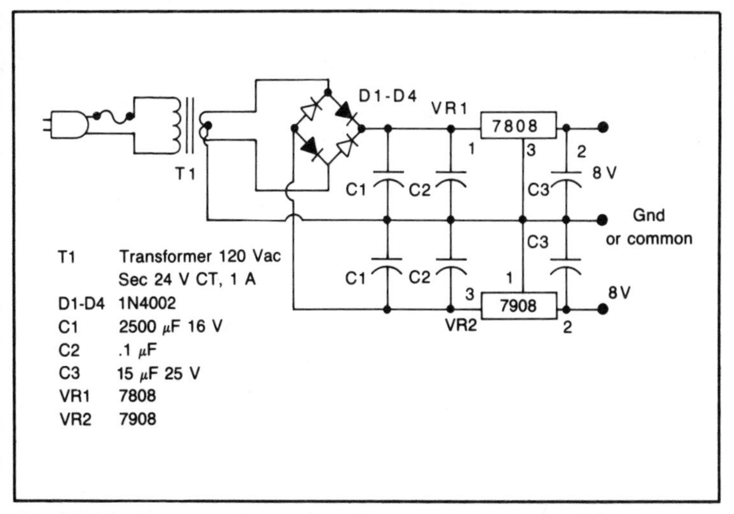

Fig. 6-2. Bipolar converter with positive and negative fixed outputs.

SINGLE ENDED CONVERTER WITH OP AMP

A fixed regulator with override option with operational amplifier control is very useful as a variable source. Figure 6-4 shows the design parameters.

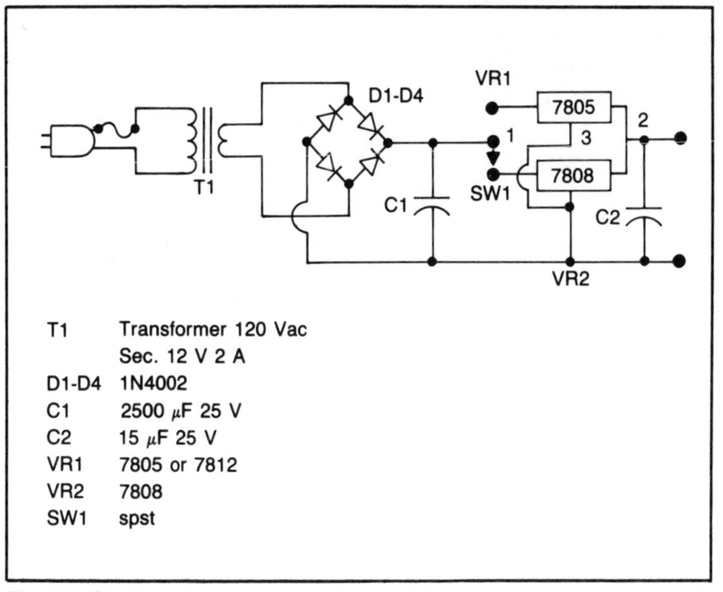

Fig. 6-3. Single-ended converter with selectable two voltage option.

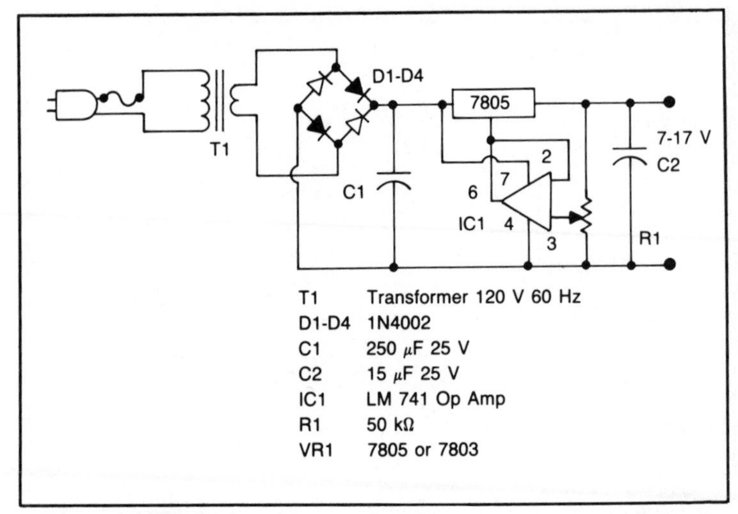

Fig. 6-4. A single-ended variable converter with an operational amplifier as the control component.

REFERENCES FOR FURTHER STUDY

The following references are good sources for energy converters.

Marston, Ray. *All About Power Supply Circuits.*, Radio-Electronics, (July-1984), page 65ff.

Motorola, Inc. *Voltage Regulator Handbook*, Theory and Practice, Phoenix, AZ, 1976, 202 pages.

National Semiconductor Corp., *Voltage Regulator Handbook*, Santa Clara, CA., 1975.

CONVERSION OF AC TO DC ENERGY

Integrated circuits operate on low voltage for the most part, therefore a transformer for ac-to-dc conversion is a step-down type. The 120 Vac at 60 hertz from the power company is stepped-down from 120 volts (rms) to 12 volts (rms) as determined by the ratio of primary to secondary turns. A simple formula (T_p: T_s = E_p: E_s) will verify this concept.

The 12 volts at the secondary of the transformer is the average of the changing frequency or rms. A view of this with an oscilloscope will show a sine wave with a peak to peak (P-P) voltage of 33.9, or (12 × 2.828). If one divides this by 2, the result is 16.96 peak voltage (PV) or (12 × 1.414). This value is important because in an ac-to-dc conversion the final dc level follows the PV not the rms. Figure 6-5 shows the relationship of the P-P to the PV.

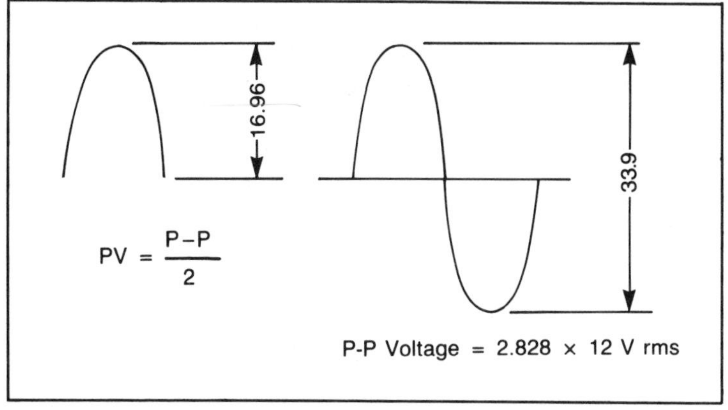

Fig. 6-5. The relationship of the PV to the P-P voltage of an ac wave of 12 V rms.

A bridge converter (rectifier) is utilized in most sources for energy conversion because of less cost. It is called a full-wave converter because each half of the sine wave appears in the output. It is easier to understand how the conversion takes place by examining what happens in one hertz of the ac wave at the bridge because subsequent frequencies are repeated. The converter is explained in terms of the dc pulse across a load before filtering.

Look at Fig. 6-6 as the first half of the sine wave is discussed. When the ac wave from the transformer is on the upward excursion of the sine wave (meaning deficiency of electrons at the top and this is sometimes called positive excursion), the bottom side of the

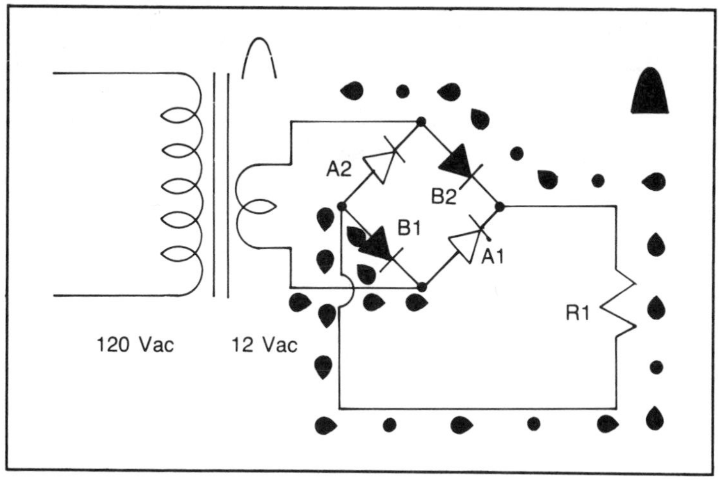

Fig. 6-6. The path of electron flow during the first half-wave of the ac input.

transformer has a surplus of electrons and they would like to find their way to the top. Electrons will flow toward diodes A1, B1 junction. Diode A1 is reverse biased (one way street) and electrons cannot cross the bridge, however diode B1 in forward biased and electrons will cross that bridge. At that point they look at diode A2 as a path to the deficient side, but that diode is reverse biased and electrons cannot cross that bridge. The only course is to go down as shown by dotted line and over to the resistive load R1 to cross it and head toward B2. Since B2 is forward biased the electrons can move toward the top. The result of the movement is a positive pulse across R1 and it is a dc pulse. One half of the sine wave is completed at this point.

Look at Fig. 6-7 as the second half of the sine wave is discussed. When the ac wave from the transformer is on the downward excursion of the sine wave, (meaning deficiency of electrons at the bottom) the topside of the transformer has a surplus of electrons and they would like to move to the deficient side. Electrons will flow toward diode B2, A2 junction. Diode B2 is reversed biased and the electrons cannot cross that bridge, however diode A2 is forward biased and they flow across that bridge. At that point they look at diode B1 as path but that diode is reversed biased and electrons cannot cross that bridge. The only course is down as shown by the dashed line and over to resistor R1, cross it and head toward A1. Since A1 is forward biased the electrons can move toward the bottom. Note that during both halves of the sine wave the electrons flowed through the resistor load R1 in the same direction and thus a positive pulse. When

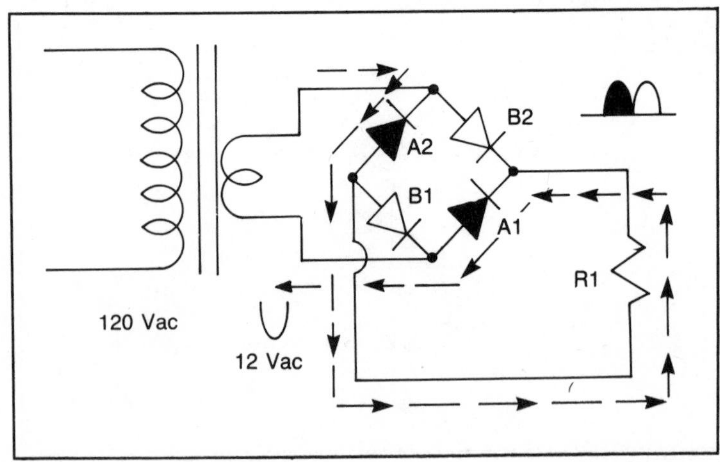

Fig. 6-7. The path of electron flow during the second half of the sine wave.

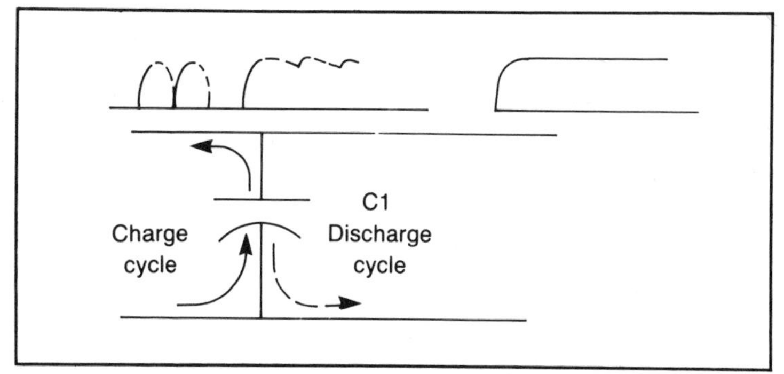

Fig. 6-8. The transition of the dc pulse to dc with ripple and eventually pure dc.

the frequency repeats at 60 hertz per second, we see 120 positive pulses all in the same direction thus converted ac-to-dc pulses.

The next step is to remove the resistor load R1 and replace it with a large capacitor. The capacitor charges on each half wave and begins to discharge when each pulse decreases. The end result is to add electrons back to the common leg by the amount the voltage trys to fall at the top and thus the pulsating dc begins to level at the top to become a steady dc level.

Figure 6-8 shows the action of the capacitor with ripple level indicated but above the zero reference. If the capacitor (tank) is large enough like 2500 μF and a three point regulator is utilized, the dc level will have a ripple that will be difficult to measure and approaching the equivalency of a battery. Note the dc level is almost the peak value of the sine wave. The slight loss is due to the voltage drop across the diodes, which is about 1.2 volts.

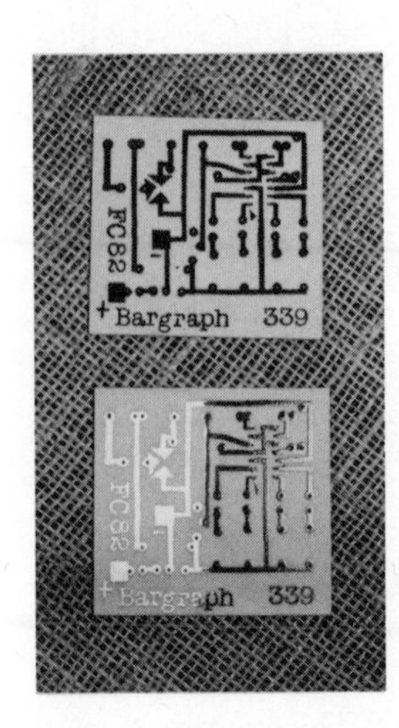

Chapter 7

Sound Control Projects

PROJECT 1
SOUND ALARM (WAILING)

Function

The wailing sound alarm is an excellent deterent against intrusion. The sound is so unusual that the intruder would be surprised.

IC1, LM555 timer, is set to operate at a rate of one pulse per second, which is monitored by D1-LED. The output is coupled through C2 to pin 13 of IC2 (one fourth of a Bilateral Switch). Pin 1 of the switch is high to the positive and pin 2 is coupled through R4 and C3 to ground by way of pin 9 and IC3 (phase-locked loop). The voltage-controlled oscillator section of IC3 will produce an oscillation that is coupled to the speaker via Q1. (See Fig. 7-1.)

The voltage-controlled oscillator will shift frequency if voltage at pin 9 changes. When IC2 switches high by IC1, C3 charges through R4 and the bilateral switch to the positive, thus a rising voltage at pin 9 causes a rising frequency. When IC2 switches off, C3 discharges through R5 to cause a falling frequency. The length of rising and falling frequency can be changed by varying C3, R5, and R4.

The beginning frequency level is controlled by C4. Decreasing C4 will set the frequency at a higher beginning point. If IC1 pulse rate is made longer by increasing the value of C1, the rising frequency will go higher before it falls.

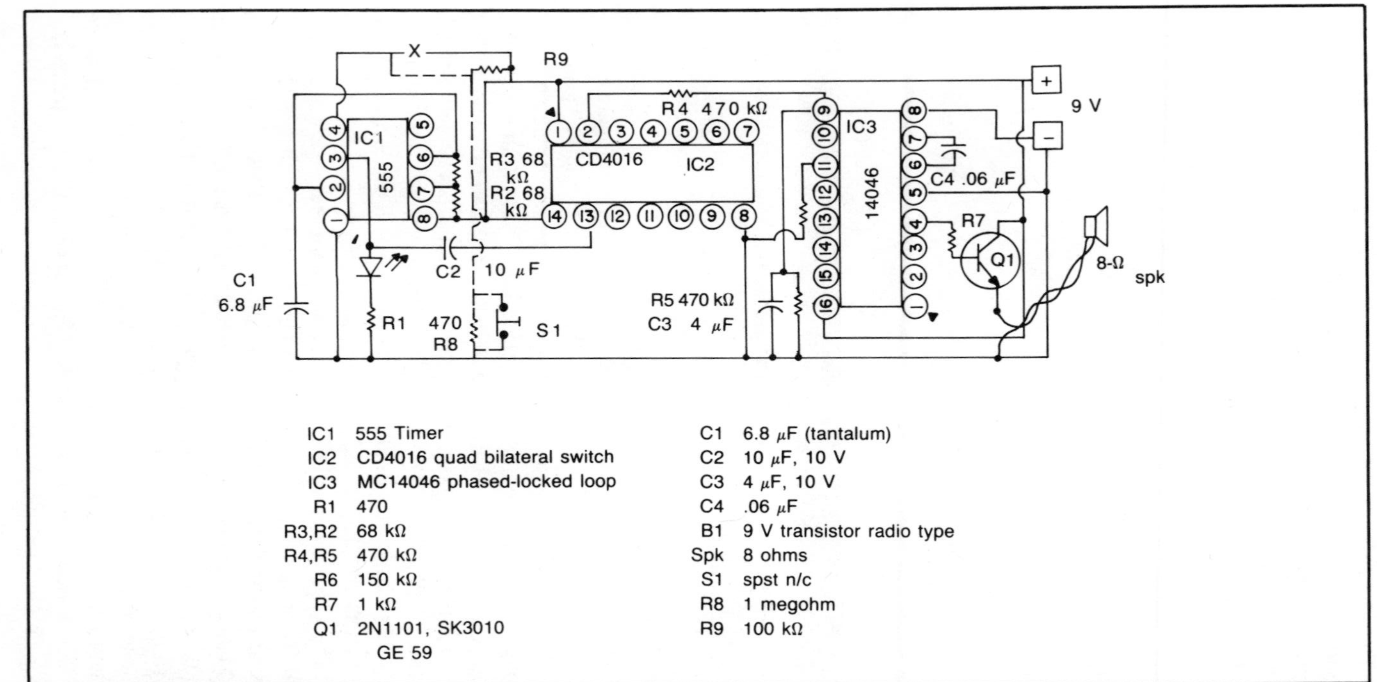

Fig. 7-1. Sound alarm (wailing).

The sound circuit by itself has possibilities, however, adding the additional components as shown by the dashed lines will make it a tripped alarm. The sound can be shut off by turning off IC1. Reset pin 4 is tied to the positive or enabling state. If it is tied low to the negative, it is off. When S1 is opened the voltage divider R8, and R9 makes pin 4 go high and the alarm is on. A piece of wire can be used in place of the switch, which sets off the alarm when broken.

Further Study

Change the value of C3 or R4 and note the change in frequency. Change the value of C1 and note the change in wailing time.

PROJECT 2
SIREN AND INTERRUPT SOUND ALARM

Function

A free running siren is useful in sports to turn on the crowd when a score is made. It can be utilized as a warning by proper trip switch or as an intrusion alarm.

The IC1, 556, is a dual 555 timer. One side (pins 1 through 6) is a slow speed oscillator that turns on and off every 8 seconds. (The other side will be explained later.) The output pin 5 is coupled through C6 to pin 13 of the quad bilateral switch (IC2). One fourth of the quad bilateral switch (pins 1 and 2) switches pin 9 of the voltage-controlled oscillator of the phase-locked loop (IC3) to the positive via R5. Pin 9 has C5 and R6 connected to the minus to complete that part of the circuit. (See Fig. 7-2.)

The voltage-controlled oscillator will cause a rising frequency with a rise in voltage at pin 9 and a falling frequency with a fall in voltage at the same pin. When the 556 (side one) goes high for approximately five seconds, C5 charges through R5 and thus a rising voltage at pin 9; output frequency at pin 4 of IC3 rises. When the 556 goes low, C5 discharges through R6 and the output frequency falls. This rising and falling frequency produces a free running siren when S1 is switched to minus. Switch S1 is connected to pin 5 of

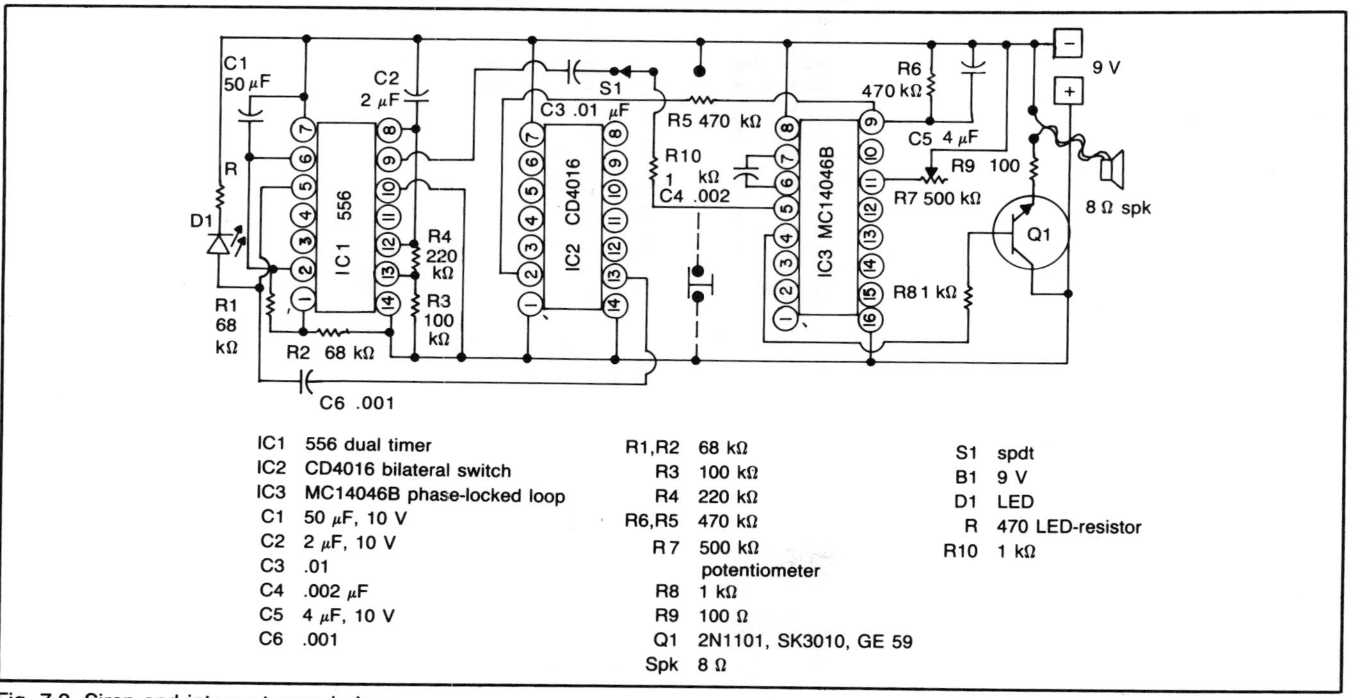

Fig. 7-2. Siren and interrupt sound alarm.

IC3, which is low enable; by switching it to minus, it permits the oscillator to operate. Note how pin 11 of IC3 is tied to minus through R7. Adjusting R7 gives a setting point for the oscillator; i.e., low, medium, or high range.

The enable low of pin 5 of IC3 provides another control function. If pin 5 is switched low the oscillator is free running; if switched high, the oscillator is shut off. Utilizing this characteristic provides a way of interrupting the rising and falling frequency. The second side of IC1 (LM556 pins 8 to 13) provides another low-frequency oscillator that is OFF and ON at 1.5 seconds with components shown in the schematic. Output at pin 9 of IC1 is coupled through C3 and S1 to enable pin 5 of IC3, thus shutting down the siren at select intervals of the output. Changing the value of C2 will change the interrupt frequency. Shorter intervals of IC3 will occur if C2 is made smaller (try a .1 μF in place of 2 μF).

The circuit can be a tripped alarm by adding switch S2 as shown in dashed line on the schematic. When closed, as shown, the alarm is off. Opening the switch will produce the sound—siren or interrupt sound.

Further Study

Try changing the value of C2 as suggested to make the interrupt changes. Add switch S2 to make a triggered alarm.

PROJECT 3
TWO-POSITION ALARM (DOORBELL)

Function

A two position alarm could be utilized for a front and back door warning or for security of some location with two entrances.

IC1 is an op amplifier in an oscillator configuration. Output from pin 6 is a feedback to input pin 2 in a bridge design (R1, R2, C3, C4) to sustain oscillation by way of ground through R3, C1, C2 and S1 or S2. Changing the values of any of these components will affect the tone output. Decreasing R3 will raise the frequency when C1 or C2 is switched to ground or minus. Oscillation will not take place unless C1 or C2 is switched to ground, however each one will produce a separate tone when switched or a third tone will be heard if both are switched at the same time or in sequence. (See Fig. 7-3.)

You may change the other components for experimental purposes, but it is recommended that you change the three variables C1, C2, and R3 rather than R1, R2, C3, or C4. Note switch S1 produces a lower tone because that leg has a lower capacitor value (C1). The third tone will discriminate better from the other two if C2 is .025 rather than .001. Diode D1 was added in series with R4 to improve tone quality and drive level.

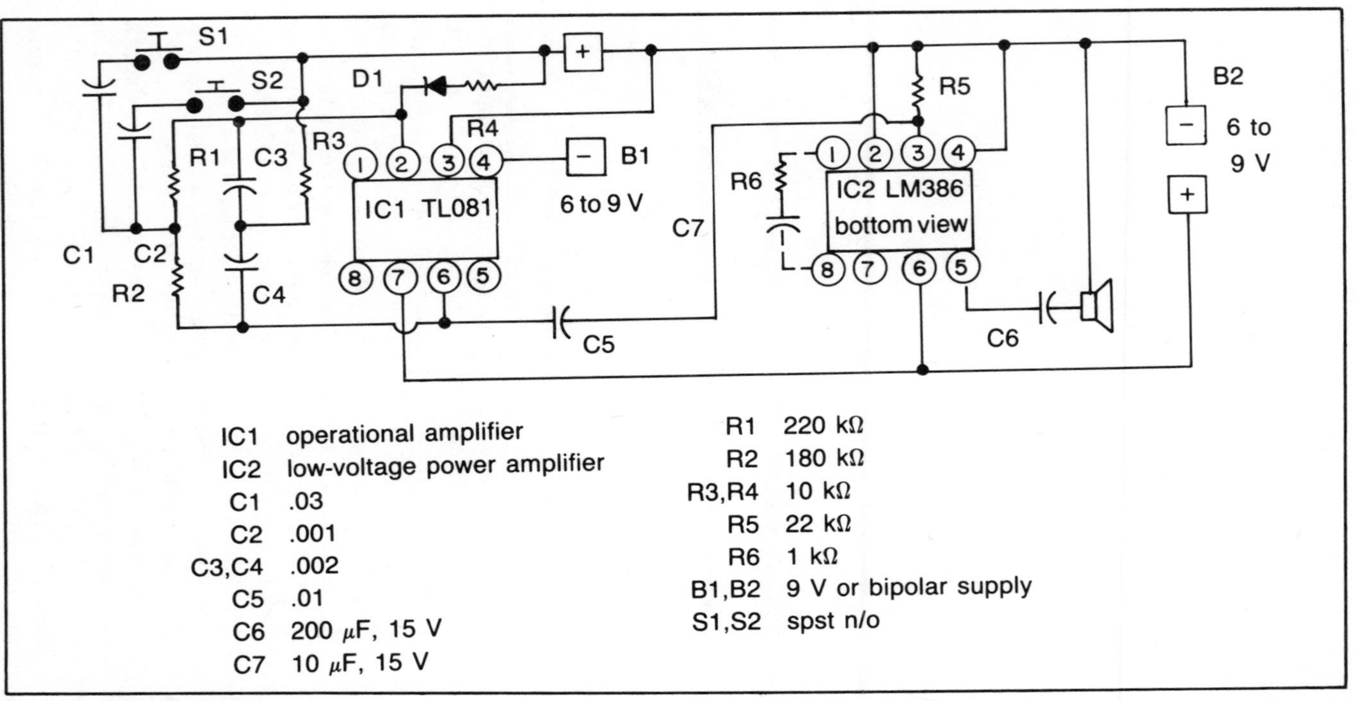

Fig. 7-3. Two position alarm (door bell).

The output pin 6 is coupled through C5 to input of IC2 (low voltage audio amplifier). This IC is adequate to drive a 12-inch speaker if one desires higher efficiency. The gain of IC2 without R6, C7 is more than adequate; with them inserted, the gain is 50 dB; with just C7 (10 μF) inserted the gain is 200.

Further Study

Change C1 or C2 to provide a greater frequency differential. Try adding another switch and another capacitor to make a three position alarm.

PROJECT 4
FOG HORN (NONINTERRUPT)

Function

The fog horn is a low tone that has an ON time controlled by a slow pulse oscillator. The tone frequency meets the specifications for a fog horn.

IC2, LM555 timer, is designed to produce a frequency of approximately 200 hertz as determined by C2, R3, and R4. The output is coupled through a 220 μF capacitor, C3, to a 12-inch speaker. A smaller speaker may be used but the larger type is more efficient at this low tone. The output could be coupled to another amplifier for greater volume but this volume is quite adequate. The ON/OFF control is accomplished through reset pin 4, which is coupled to IC1, pin 4. (See Fig. 7-4.)

IC1, CD4001-quad 2-input NOR gate, has two gates cross-coupled to become an oscillator. The oscillator frequency of IC1 is 1 pulse per 1.3 seconds or the tone is ON for that length of time. The output is almost symmetrical thus the OFF time is approximately the same. The pulse rate can be changed by varying components R1, R2, and C1. Decreasing R2 will decrease the ON time while increasing R2 will lengthen the ON time. The output is not symmetrical with this kind of change, therefore the OFF time was shorter than the ON time.

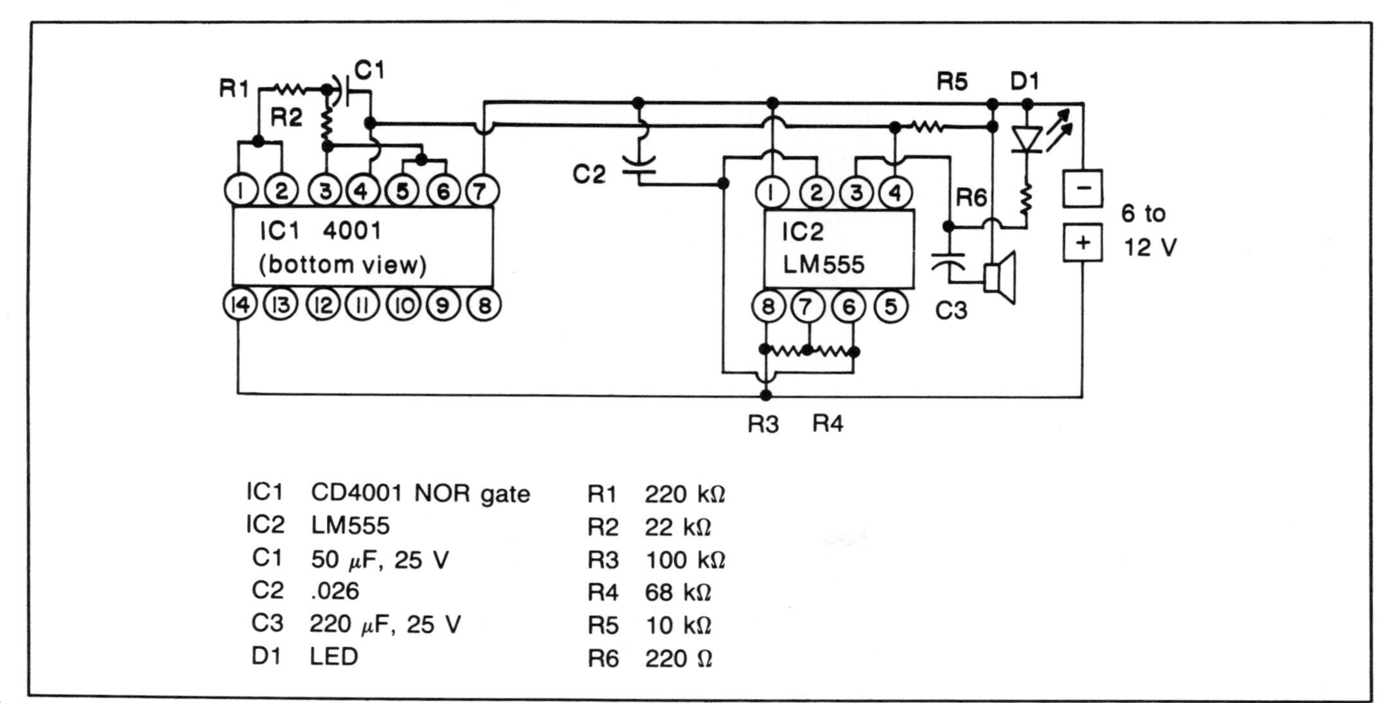

Fig. 7-4. Fog horn (noninterrupt).

The LED in the circuit is not necessary for the circuit to perform. It was added to monitor the time visually.

Further Study

Change the value of R2 or C1 to change the length of OFF and ON time. You may wish to check elsewhere in this book for a bridge amplifier that could be added for greater volume. (See Projects 16 and 19.)

PROJECT 5
FOG HORN (INTERRUPT)

Function

This circuit produces sound like a true fog horn in that there are three blasts of tone and then a delay off before repeating. The number of blasts and length of time off as well as the tone are programmable by changing select components.

IC2, LM555 timer, is a low-tone oscillator with tone controlled or programmed by changing the value of C2. A sound between 250-525 hertz is recommended for inland waters. If C2 is .022, the tone produced is 250 hertz. The output is coupled to a speaker through C3. LED D1 is optional and was placed into the circuit for visually monitoring time ON. (See Fig. 7-5.)

IC1 is a CD4011, quad 2-input NAND gate, with two gates cross-coupled by components R1, R2, and C1 to become a low-pulse oscillator of 1.3 seconds ON duration. The output pin 4 is coupled to pin 4 of LM555 to determine the length of ON time for the tone oscillator. IC1 is enabled by pin 1; when it is high or ON, the IC1 can oscillate at a low pulse rate to control the length of one blast of the tone. When pin 1 is held low, the output is low and the tone is shut off. It is possible through pin 1 to permit a certain number of tone blasts and then shut off for a period of time before the tone blasts are repeated.

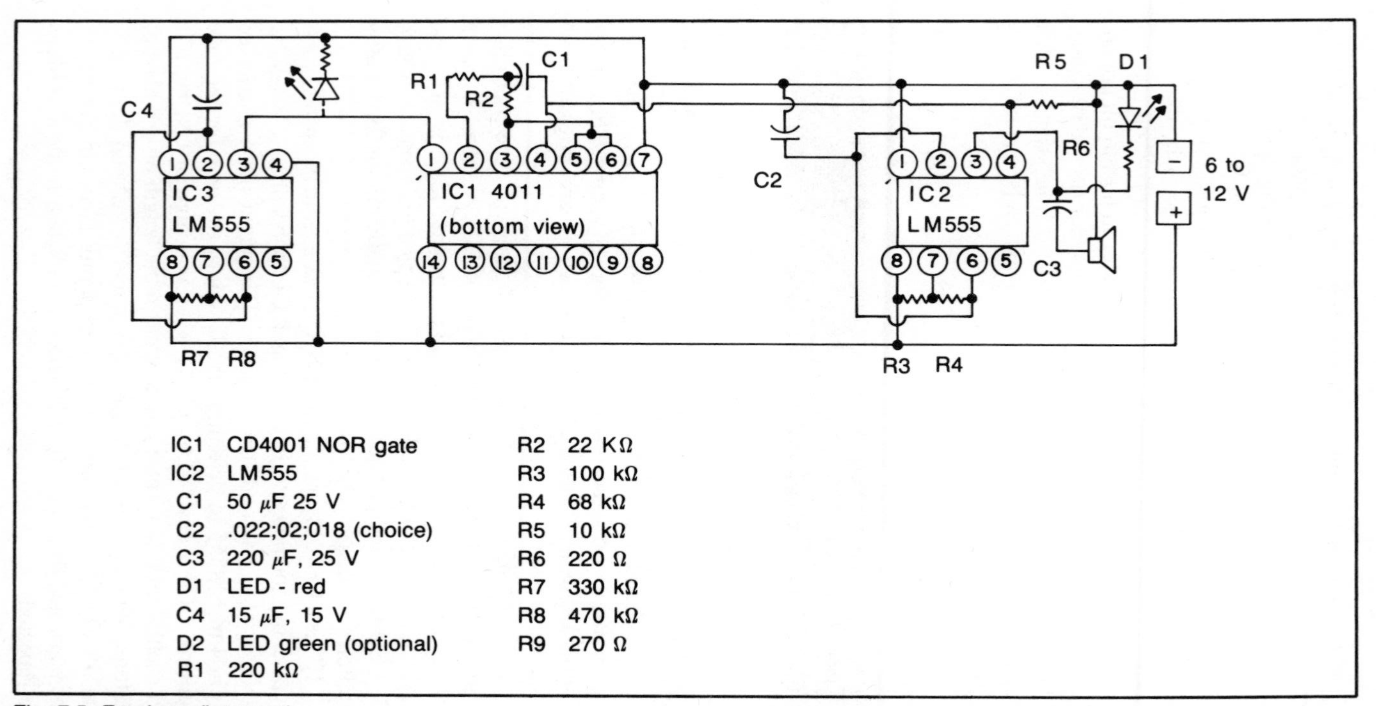

Fig. 7-5. Fog horn (interrupt).

IC3 is another low-pulse oscillator through R7, R8, and C4. By making R7 and R8 large and holding C4 at proper value, it is possible to control the number of tone blasts between shutdown periods. The output pin 3 is coupled to the gate enable pin 1 of IC1. It is easier to control by programming C4 rather than changing R7, or R8. When C4 is 10 μF one gets two blasts of tone. At 20 μF there are four blasts of tone, and at 15 μF there are three blasts of tone.

LED D2 (Green) and the 270-ohm resistor in the circuit are optional. The LED is in the circuit to visually monitor ON and OFF time.

Note: The LM555, tone generator drove a 12-inch speaker with adequate volume. The output could be coupled to an IC amplifier such as LN 386 or LN 384.

Further Study

Look in this book to find a circuit in which a LN 384, or LN 386 amplifier is utilized and adapt it to produce a louder sound. Increase the value of C2 and note tone change or increase the value of C4 to increase time OFF.

PROJECT 6
SYNTHESIZED SOUND (THEATRICAL)

Function

The synthesizing of two sounds can produce some interesting effects such as train whistle, old car horn, wind blowing, noise of the night, and others. Such sounds have use in theatrical productions. Weird sounds for Halloween also have possibilities with this circuit.

IC1, CD4011, quad 2-input NAND gate, has two gates cross-coupled to become two separate audio oscillators. Components R4, R5, and C5 on one side and R1, R2, and C1 on the other side, determine the sound outputs. Changing C5 and C1 will change the outputs from pins 4 and 10. If one utilizes a capacitor substitution box for each of these capacitors by connecting the boxes in parallel, a variety of sounds can be accomplished as one switches capacitors. A third sound effect can be produced by varying the voltage input on a swing in voltage from 5 to 9 Vdc. This will add a wail to the synthesized sound. (See Fig. 7-6.)

The outputs from pins 4 and 10 of IC1 are coupled through C2 and C6 to IC2. IC2 acts as a buffer (go between) to prevent loading of the outputs. The outputs from IC2, pins 3 and 11 are coupled to R6 as a common load. The two signals are synthesized at this point

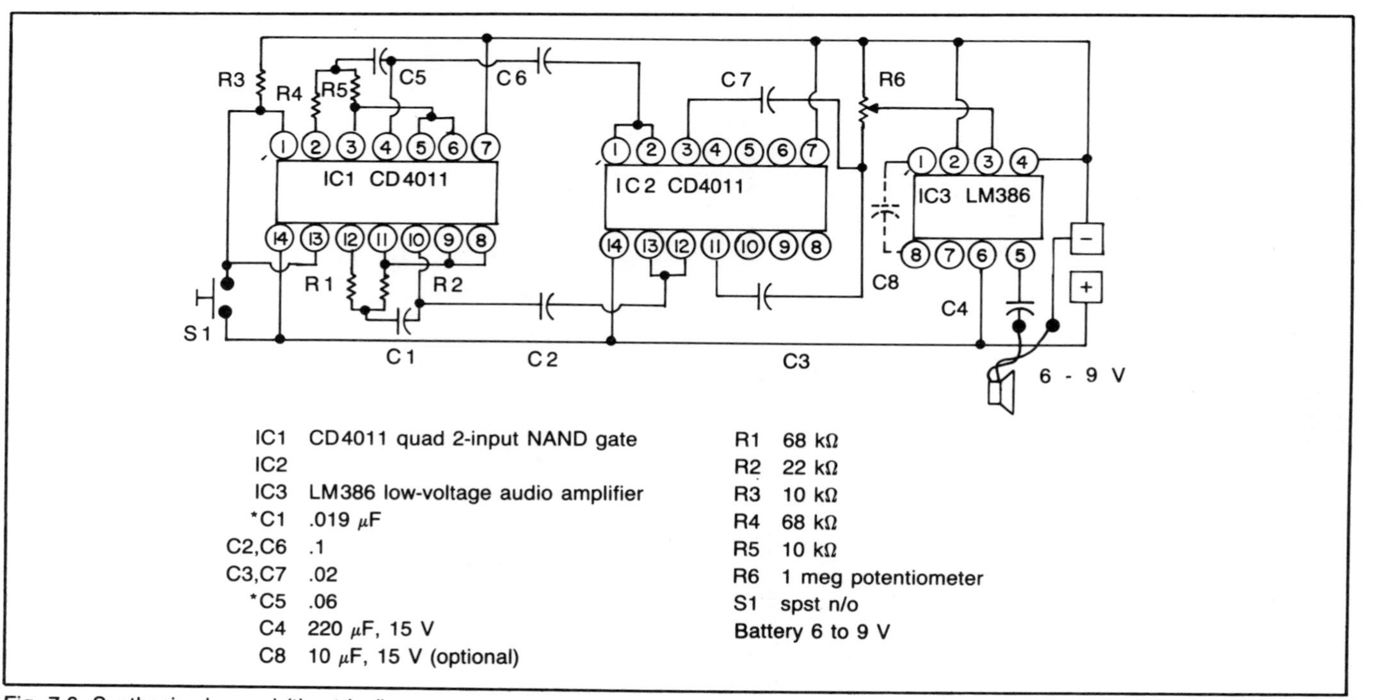

IC1	CD4011 quad 2-input NAND gate
IC2	
IC3	LM386 low-voltage audio amplifier
*C1	.019 μF
C2,C6	.1
C3,C7	.02
*C5	.06
C4	220 μF, 15 V
C8	10 μF, 15 V (optional)

R1	68 kΩ
R2	22 kΩ
R3	10 kΩ
R4	68 kΩ
R5	10 kΩ
R6	1 meg potentiometer
S1	spst n/o
Battery 6 to 9 V	

Fig. 7-6. Synthesized sound (theatrical).

and R6 sets the volume level to drive IC3 (LM386 audio amplifier).

IC3, LM386, is a straightforward audio amplifier.Capacitor C8 is optional and can be added to give a power factor of 200. Without C8 the power factor output is 20. Series resistor/capacitor combinations can produce other power factor levels.

Further Study

Change the values of C5 and C1 and note the changes that can be made in the synthesized sound. Try a variable power pack and power IC1 and IC2 from this pack and keep IC2 audio at a fixed voltage. Swing the voltage between 5 and 9 volts to produce weird sound effects.

PROJECT 7
SYNTHESIZED SOUND (REPETITIVE)

Function

The synthesized sound in repetitive mode is an excellent alarm device for getting attention and it can be triggered by an appropriate transducer in place of S1. The circuit is similar to the *synthesized sound (theatrical)* in Project 6 except for a change in layout of IC2. In this part of the circuit, IC2 pins 1 through 6 are utilized as the buffer side. Both frequency inputs are coupled in and out via C6, C2, C7, and C3. The buffer pins 13, 12, and 11 were moved to the other side to make room for a third cross-coupled oscillator. The components R7, R8, and C8 were made large to produce a slow pulse rate (frequency) of 1 hz per 6 seconds or an ON and OFF time of 3 seconds each. Increasing or decreasing the value of C8 will change the duration of the repetitiveness. The output pin 10 is coupled directly to pins 1 and 13 that enable the oscillators to run when they go positive. The LED, D1, and R9 hold the enabled pins 1 and 13 to ground. In addition, the LED monitors the ON time as the LED lights when enable is high or positive. (See Fig. 7-7.)

IC2 is enabled through pin 13. When it is high the sound is ON. Switch S1 and R3 provide for enabling or disabling. If S1 is replaced

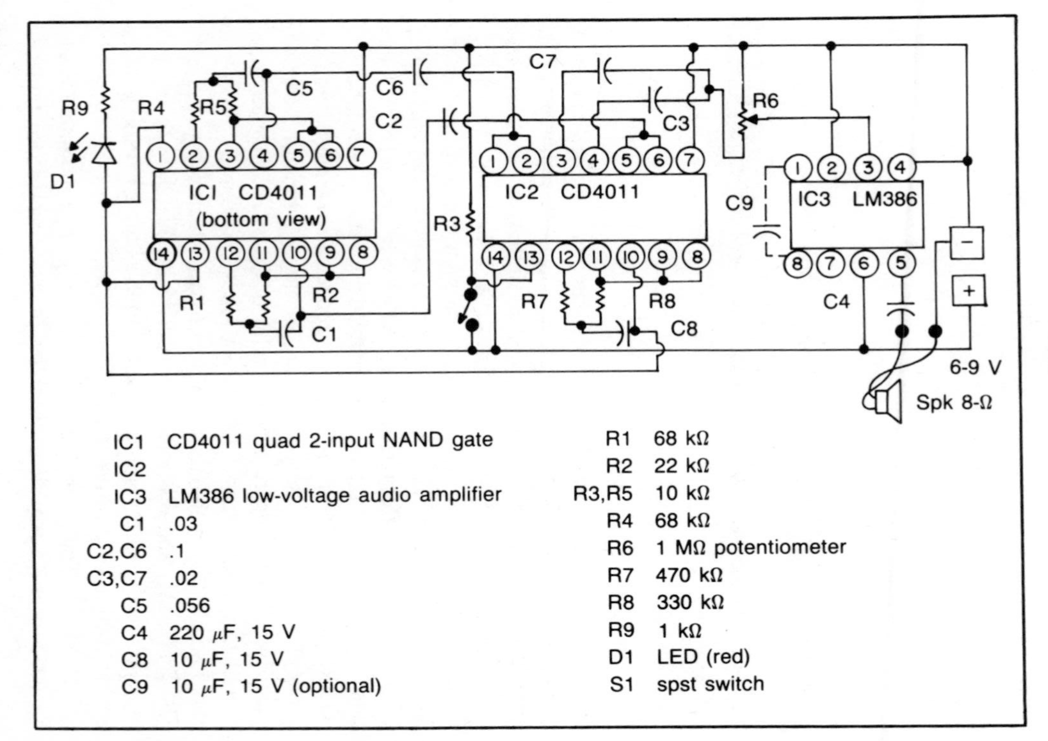

IC1	CD4011 quad 2-input NAND gate	R1	68 kΩ
IC2		R2	22 kΩ
IC3	LM386 low-voltage audio amplifier	R3,R5	10 kΩ
C1	.03	R4	68 kΩ
C2,C6	.1	R6	1 MΩ potentiometer
C3,C7	.02	R7	470 kΩ
C5	.056	R8	330 kΩ
C4	220 μF, 15 V	R9	1 kΩ
C8	10 μF, 15 V	D1	LED (red)
C9	10 μF, 15 V (optional)	S1	spst switch

Fig. 7-7. Synthesized sound (repetitive).

with a spst n/c switch and it is held open by force (similar to a car door light switch) the unit becomes a triggered alarm when the door is opened or the force is removed. Many trucks have a similar device that signals when the vehicle is in reverse and about to move backward (use a back-up light for connection in this regard has a possibility). Look elsewhere in this book for a circuit called Backup Beeper (Project 31).

Further Study

What affect does changing the value of C8 have on the repetitive rate? Can you change the harmony of the synthesized sound by lowering or raising the value of C5?

PROJECT 8
ALTERNATE TONE ALARM

Function

A two-tone generator that is alternately switched ON provides a high/low output as might be heard from a traffic vehicle like a police car or ambulance.

IC1, CD4011, quad 2-input NAND gate is a two tone oscillator in which each side, pins 1 through 7 and 8 through 13 set the tone frequencies. Changing the values of C2 and C1 determines the high/low tones. The output frequencies are coupled to IC2, CD4011, of which one side (pins 1 through 6) acts as a buffer. The buffer is necessary to prevent loading on the outputs that would occur if one tried to go directly to the LM386 amplifier. The other side of IC2, pins 8 through 13, is a slow pulse oscillator of approximately 8 hertz per second. The output at pin 10 is connected to IC4 as a clock. (See Fig. 7-8.)

IC4, CD4027, is a dual J-K master-slave flip-flop that is wired to perform as a toggle switch in which Q1 pin 15, and Q1 (NOT) pin 14, go high and low alternately (flip-flop). The clock input from IC2 pin 10 is connected to pin 13 of IC4, and the outputs at pins 15 and 14 changes the flip/flop state with each positive pulse transition. The CD4027 functions in toggle mode when the set and reset inputs,

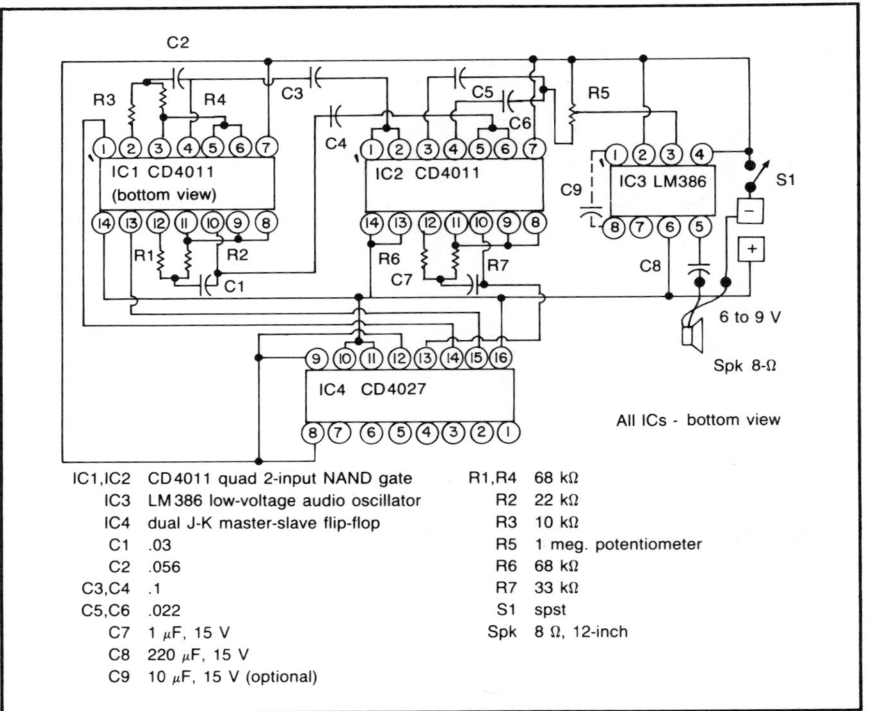

Fig. 7-8. Alternate tone alarm.

pins 9 and 12, are held low or grounded. Also, J-K inputs, pins 10 and 11, must be held high or to the positive. The outputs Q1 and Q1 (NOT), pins 15 and 14 are connected to pins 13 and 1 respectively of IC1 that enables or disables. Thus, each tone oscillator is turned on and off alternately. IC3 is a straightforward low-voltage audio amplifier.

Further Study

Try changing the tones of the oscillators by changing the values of C1 and C5. How about changing the toggle rate by changing the value of C8. Why not see what happens when set/reset is changed to high. Can you add an LED and a 470-ohm resistor to monitor the clock rate visually?

PROJECT 9
TRUCK HORN (AIR SIMULATED)

Function

A truck horn of air-blast type produces sound with authority to attract attention. Twin trumpet horns can be purchased for an automobile but they require a compressor and associated electrical circuit. The circuit shown here produces this type of sound electronically.

IC1, LM556 is a two tone oscillator that drives IC2, a Dual 4 Watt Amplifier. The LM556, pins 1 through 6 are the higher tone source of about 200 hertz. The other side, pins 8 through 13 is the low tone source of about 140 hertz. Changing the values of C3, C2 will change the tones of the oscillators. The outputs of IC1, pins 9 and 5 are connected to separate potentiometers to control volume and balance. (See Fig. 7-9.)

Each side of IC2, LM378, produces four watts of audio that is connected to two 8-ohm speakers through C7 and C8 respectively. The speakers should be indoor/outdoor weather resistant 5-inch horn types. The operating current was measured at 400 mA or approximately 5-watts rms from the supply.

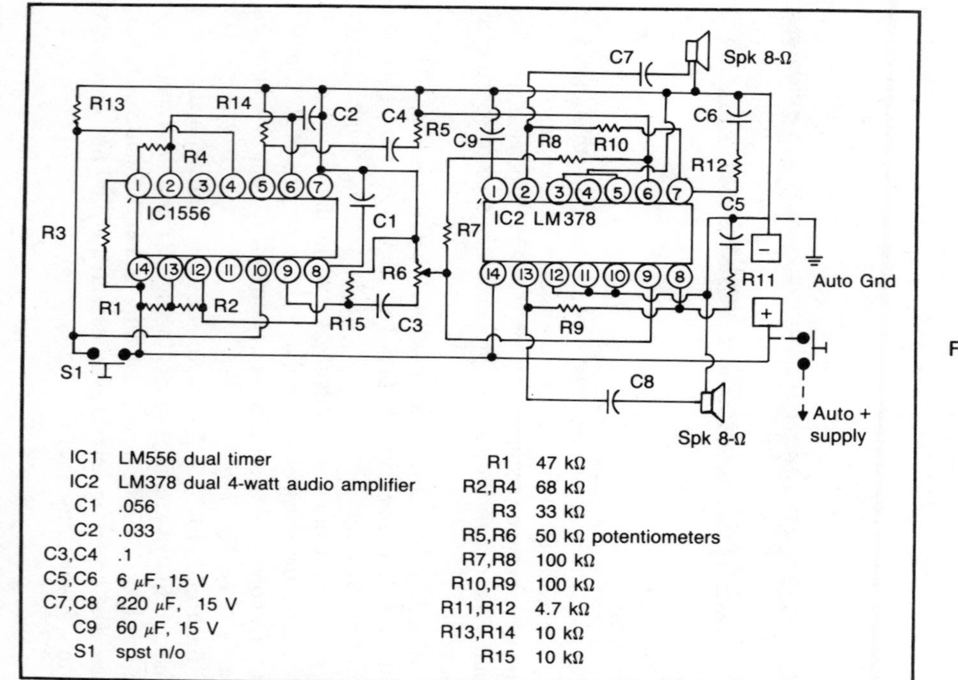

Fig. 7-9. Truck horn (air type simulated).

Further Study

Change values of C1 and C2 to note change in frequency. Use a jumper wire in place of S1 and try control as shown in dashed lines to simulate with the power supply as though the supply was from the truck battery.

PROJECT 10
TWO-STATION INTERCOM

Function

The circuit is designed to provide separate amplifiers for each station rather than a single amplifier system. This reduces the number of wires to three for interconnection. It also provides further distance between stations.

IC1 and IC2 are low-voltage audio amplifiers. Each operate as separate entites with switching as the control factor. The LM386 provides a gain of 20 with C7 and C8 removed from the circuit. When they are in the circuit, the gain is 200. Other gain levels can be achieved by adding a resistor and a capacitor in series between pin 1 and 8 such as 1 kΩ and 10 μF, which produces a gain of approximately 150. The gain of each stage can be adjusted independently depending upon the drive capability of the speakers when sending. The speakers should be of the highly-flexible cone type in order that the voice level can be of normal volume at 3-foot distance. (See Fig. 7-10.)

The switches should be spring loaded, spdt, normally closed at one position. The switches should be connected so the close position side is in the listen mode or speaker in receive. One pushes the switch and holds to talk; always returning to listening mode when released.

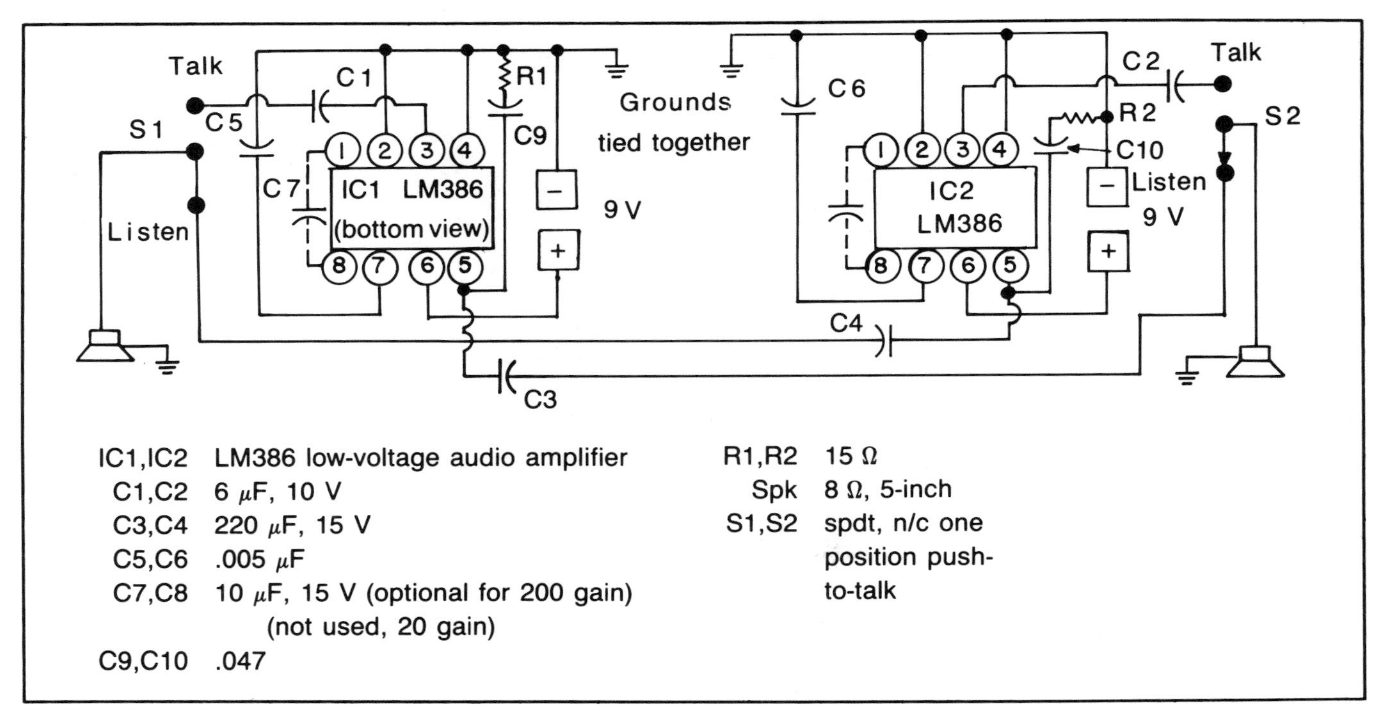

Fig. 7-10. Two station intercom.

Components R1, C9 and R2, C10 are in the output circuits to improve quality. Oscillations will occur at the bottom end of the negative excursions of the wave without these components.

Note: The third wire is a ground return. This can be another wire or it can be a ground return if a good metallic ground is available.

Further Study

Change the values of C7 or C8 and note any change in output sound. Don't forget that a resistive value may be added in series with any capacitor change to affect output level. Could you connect another LM386 to produce a third station?

PROJECT 11
MASTER INTERCOM WITH REMOTE STATIONS

Function

The Master Intercom is a small unit with large capability. The circuit shown was capable of good communications up to 40 feet of line without a line amplifier. Longer distances might require a line amplifier near the remote stations to compensate for line loss. Voice sound from a distance of six feet from a 12-inch speaker at the base station provided good readability at the remote stations.

IC1, LF351, FET op amplifier is utilized as a preamplifier with gain determined by the R3/R4 ratio or 1000 with 1 megohm/1 kilohm resistors. An op amplifier requires bipolar voltage for maximum gain. Also, the offset voltage at output pin 6 must be set to zero by null pins 1 and 5 and resistor R3. Use a digital voltmeter set between output pin 6 and common then adjust R3 to get as close as possible to zero voltage. The drive level voltage is set by potentiometer R5. Switches S1 and S2 must be spst, n/c one position, spring loaded or push-to-talk types. The normally closed side must be in the IC2 amplifier side for listening mode. Note all speakers are switchable to the op amplifier for talk mode. (See Fig. 7-11.)

IC2 is a dual 2-watt audio amplifier with one side for the base station voice transmission and the other side for the return from the

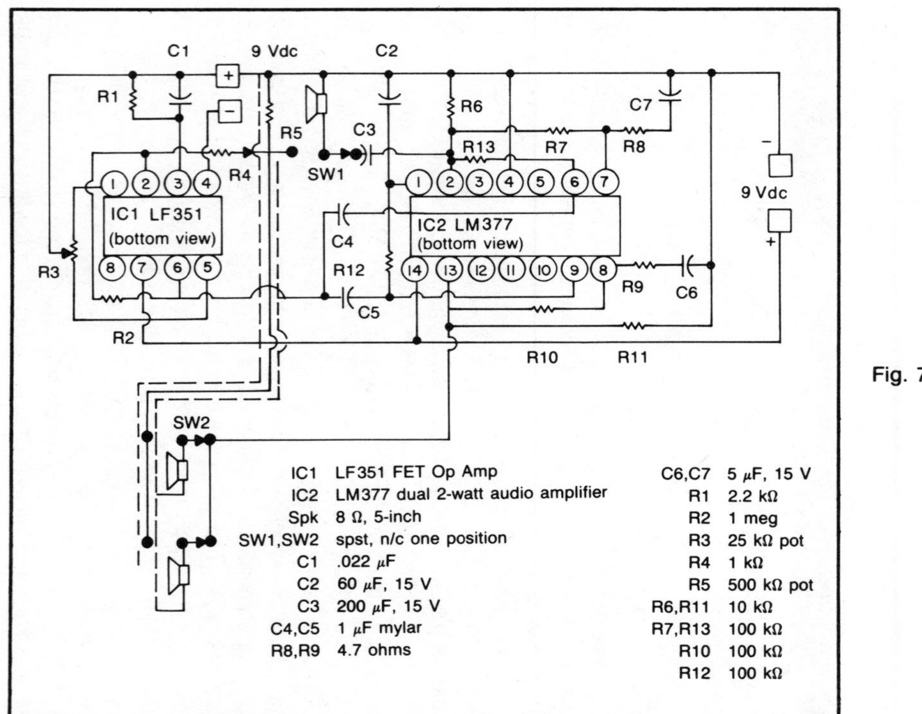

Fig. 7-11. Master intercom with remote stations.

remote stations. The output for IC1 feeds both amplifiers at the same time, therefore R6 and R11 are in the outputs as loads during the period when one is talking and that speaker is unloaded to the amplifier output. Components R8, R9, C6, C7, R7, R10, R13, and R12 are in compensating networks for stability and gain.

Further Study

Try different distances from speakers for voice pickup to test the sensitivity and proper setting of R5. Be sure the shielded cable is grounded. Remember if the remote stations are long distances from the master station, a line amplifier similar to the LF351 will need to be installed at the remote stations.

PROJECT 12
BOAT HAILER OR BULL HORN

Function

The ability to be heard in a crowd or directional communications is possible with a boat hailer or bull horn with a narrow cone of dispersion.

The input to the boat hailer begins with IC1, LM386, (a low-voltage audio amplifier) pin 3 by way of C3 and R1. The potentiometer R1 sets the drive level or volume. Gain of the LM386, which is acting as a driver stage can be set from 20 to 200 depending on the connection of components between pins 1 and 8. If C2 is 10 μF, the gain is 200, which is about right for this circuit depending on the drive level of the microphone in use. A resistor added in series with the 10 μF capacitor will reduce the gain. For example, a 1.2 kΩ resistor in series with the 10 μF capacitor reduced the gain to 50. (See Fig. 7-12.)

The output from IC1 is from pin 5 to IC2, pins 6 and 9 via C4 and C5. The dual 2-watt amplifier is connected in parallel to produce about 4 watts of audio power. Frequency stability is obtained through R2, R4, and C10 on one side and corresponding R6, R5, and C9 on the other. The output pins 2 and 13 are coupled to the speaker through C6 and C7. All input leads should be kept as short as possible and

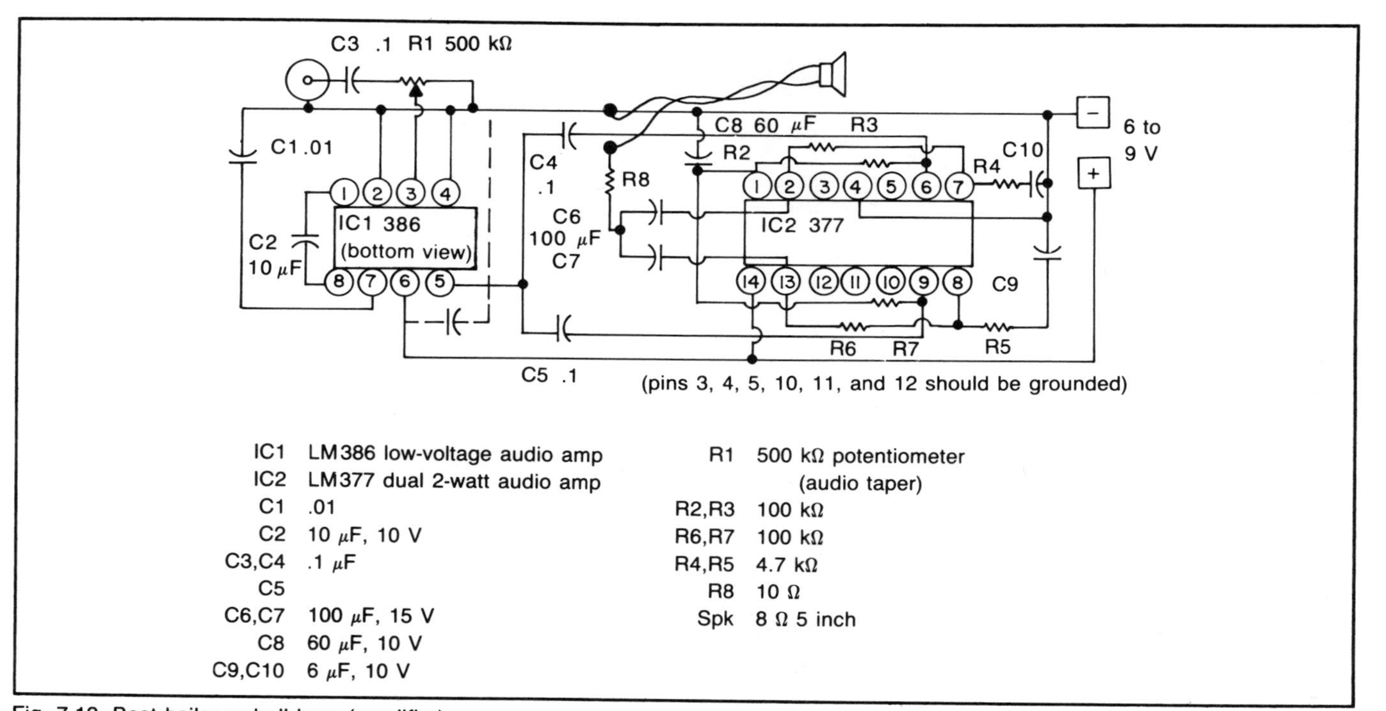

Fig. 7-12. Boat hailer or bull horn (amplifier).

use shielded cable to prevent extraneous pickup and feedback. The input to the first stage is particularly vulnerable. The LM386 has tendency to self oscillate in proto-board layout and sometimes in an etched circuit board. Decoupling of the dc with a .05 capacitor between pin 6 and ground is essential to avoid this problem.

A high-output microphone is advisable, although good results were produced with a crystal microphone with 200 mV P-P drive. Remember the speaker must be ahead of the microphone and the cone pointed away to prevent feedback.

Further Study

Remove C2 from the LM386 circuit and note the drop in loudness. Try a 5 μF replacement and note the control of maximum loudness. Scope the signal input at the microphone and at pin 5 and note voltage gain. If you use a microphone instead of an audio generator you will have to hum a tone at constant frequency and volume for readability.

PROJECT 13
STEREO AMPLIFIER WITH PREAMPLIFIER

Function

A small stereo amplifier is often utilized for a children's record player. Four watts of power for each channel is excellent for this purpose.

The preamplifier, IC1, requires two separate voltages with two batteries or a bipolar supply. The LM1458 operational amplifier requires (1) a positive voltage at pin 8 and common as negative as one set of voltages and (2) a negative voltage at pin 4 and common as positive as the other set of voltages. The gain is approximately 100 or calculated by dividing R4/R2 (or in this instance 1 megohm/10 kilohm) for each side of the dual operational amplifier. This gain is adjustable by increasing the value of R3 and R4 since gain is determined by the feedback resistance divided by the input resistance. The signals are applied to the inverting input, pins 2 and 6, while the noninverting inputs, pins 3 and 5, are tied to ground or common. No effort was made to offset balance the outputs and one side had higher gain than the other, however the signals can be balanced at the next stage. (See Fig. 7-13.)

The signals from IC1 are coupled from pins 7 and 2 through C6 and C3 to dividing resistors R10 and R6 respectively. These should

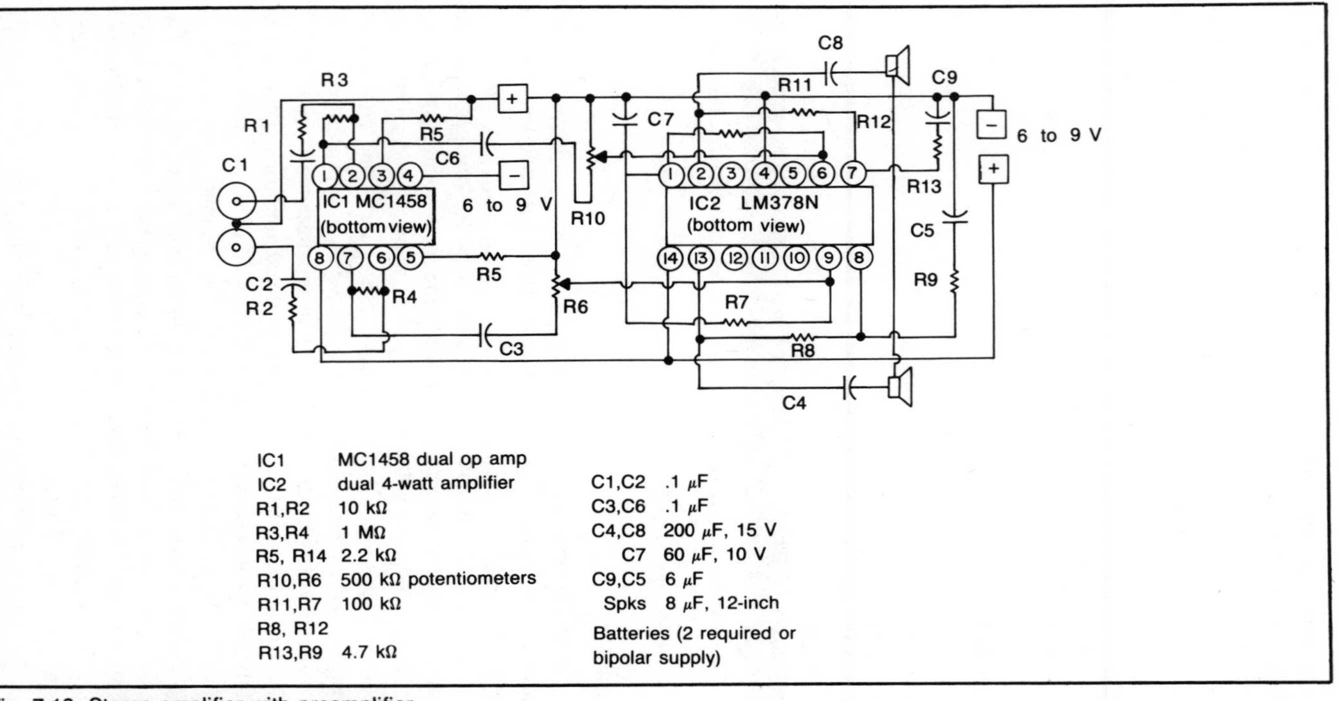

Fig. 7-13. Stereo amplifier with preamplifier.

be adjusted for correct volume and balance. The dual 4-watt amplifier, IC2, has the same frequency compensating connections and stabilizing components as were explained in the description of circuit function for the LN 377. The ICs are similar except for power output. Each side of the LM378 (IC2) is a 4-watt amplifier and has no difficulty in driving a 12-inch speaker. In fact, you can omit the preamplifier if the driving voltage from the microphone or cartridge is 400 millivolts P-P. Only 4 millivolts is necessary with the preamplifier at 100 gain.

Further Study

Check for cross-sound in the other channel with signal injected in the other side. If the signal is inaudible, use an oscilloscope to make this check. Use an audio generator and inject a 1 kHz signal at the top of R6 and R10 and note the sound. Without changing the volume level, move the generator back to the preamplifier and note the loudness.

PROJECT 14
TELEPHONE AMPLIFIER

Function

The telephone amplifier provides greater opportunity for the family to participate in the call from a loved one by the use of amplified conversation. No tampering with the phone is necessary because the pickup is inductive with no electrical connection.

Q1 transistor performs as a preamplifier with a gain of about 75 or 4 millivolt input to produce 300 millivolt P-P output. Q1 in a waveform check had a beta of about 70. The output signal is coupled through C3 to R7 as a voltage divider to control volume or drive level to the next stage. If a higher gain transistor is used, change the value of the R9 bias resistor (upward) to produce a Q point of one-half the supply voltage. Q point is measured from minus to collector of Q1. (See Fig. 7-14.)

IC1 is a dual 2-watt amplifier connected in cascade, one side (pins 1 through 7) acts as a driver while the other side (pins 8 through 13) is the final amplifier. Compensation and balance is accomplished by components R1, R2, R3, R4, R5, R6, C4, C6, and C7. Pins 3, 4, 5, 10, 11, and 12 should be grounded to minus. Pin 4 has to be grounded because it is the minus input to the IC. Keep all connections to inputs as short as possible in layout to avoid extraneous fre-

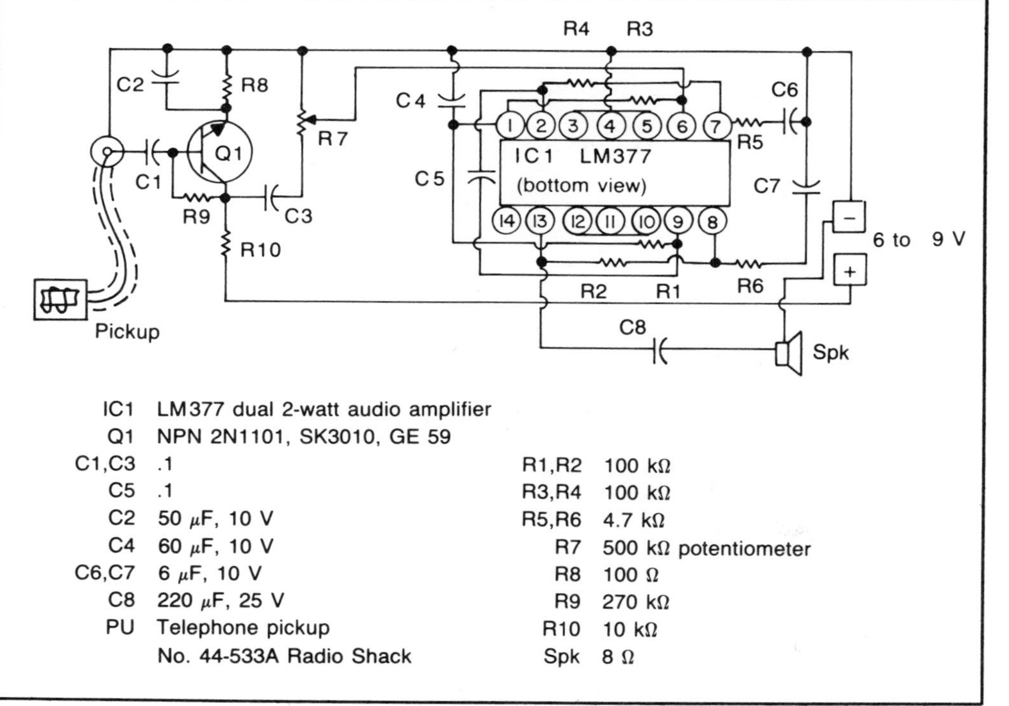

Fig. 7-14. Telephone amplifier.

quencies. There might be a tendency to overdrive so adjust the volume to a nondistorted signal at the output. The pickup has a suction cup that should be connected to the receiver of the telephone.

Further Study

Try another transistor of higher beta and make adjustments for good linearity. The key is to readjust the bias to get a Q point of one-half the supply voltage. Remember the grounded pins are connected internally and grounding helps to heatsink the IC.

PROJECT 15
8-WATT AUDIO AMPLIFIER (WITH PREAMPLIFIER)

Function

The 8-watt power amplifier with preamplifier would be valuable as a room public address device. It would be an excellent portable public address system for speakers or organizations like Lions, Kiwanis, Rotary, and the like. The high performance FET op amplifier has excellent gain and low noise for a low-level microphone or phono cartridge.

IC1, LF351, FET operational amplifier requires a bipolar voltage at pins 4 and 7 with a common ground to provide maximum gain. The gain is calculated by dividing R2 by R1. The gain in this circuit is approximately 500 or 470 k/1 k. All operational amplifiers exhibit an off-set voltage at output pin 6 with the input open. this must be set to near zero for good linearity and gain. The zero set balance can be achieved through pins 1 and 5 via R3. One places a voltmeter between pin 6 and ground and then adjusts R3 to produce a zero voltage. Once achieved the potentiometer can be measured for resistance each way from center tap and then replaced with fixed resistors. A 20 millivolt P-P input signal will drive the final amplifier. (See Fig. 7-15.)

The components R6, R7, R8, and C3 provides a tone control for

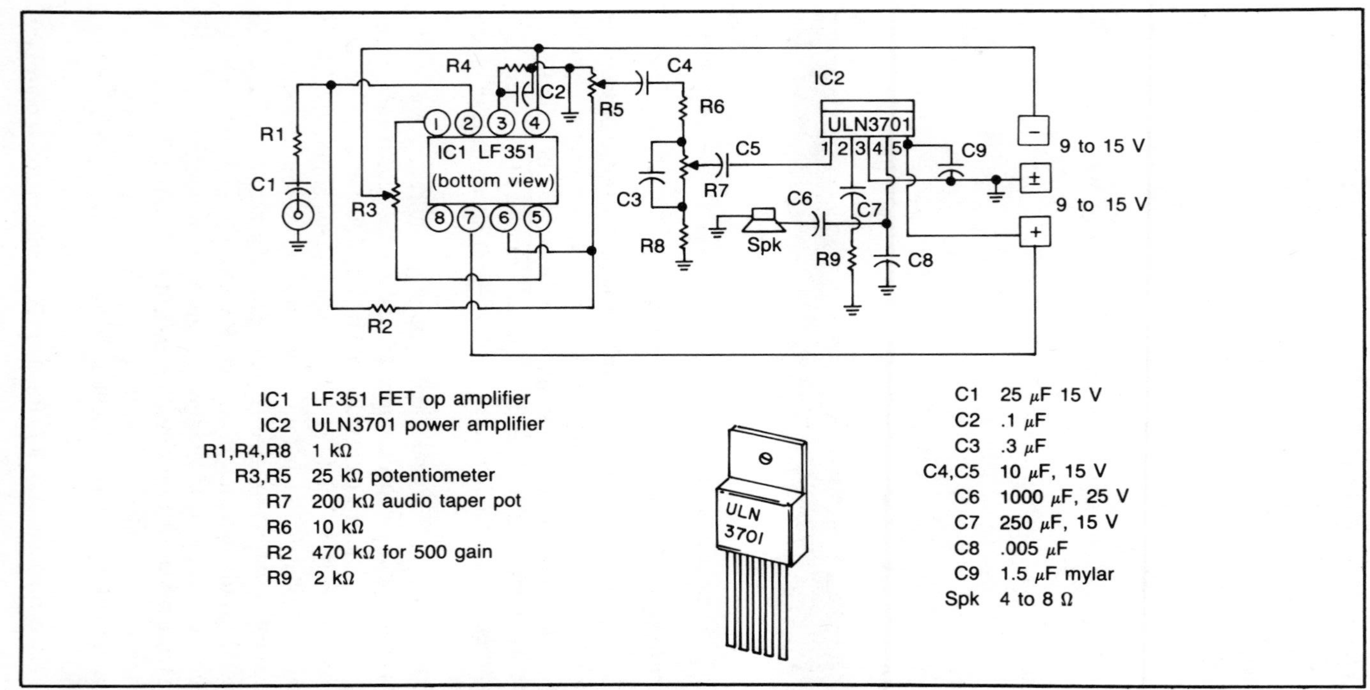

Fig. 7-15. 8-Watt audio power amplifier with Preamplifier.

bass boost. The bass boost at 50% can be calculated by $F_2 = 1/(6.28\ R7\ C3)$ (which is 500 hertz) or $f_2 = 1/(6.28 \times .2 \times .0015)$.

The input signal to IC2 (power amplifier) is connected to the noninverting input. The output pin 4 is connected to the speaker via C6 which should be 1000 μF or larger for full power capability. Component C9 is the dc decoupling capacitor and C8 is for parasitic suppression. Supply voltage can be from 9 to 15 volts without altering component values because the internal circuit keeps the Q point at pin 4 to 1/2 the supply voltage for good linearity. The standby current from the power amplifier is 100 mA while the op amplifier standby current is 2 mA.

Further Study

Change the value of C3 and note the difference in bass response. Change the value of R2 to 1 megohm and note the difference in gain by scoping the input and output pin 6 of IC1.

PROJECT 16
8-WATTS PER CHANNEL STEREO AMPLIFIER (WITH PREAMPLIFIER)

Function

A mini-brute stereo amplifier would be useful as a portable amplifier for a record player or public address system for a wide hall.

IC1, LF353, is a dual-input FET operational amplifier with a gain of 200, (V_g = R3/R2). The input drive and balance is accomplished through potentiometer R1 for each channel. A bipolar supply is necessary for the operational amplifier for good gain and linearity of the output. (See Fig. 7-16.)

The signal to IC2 and IC3 goes to pin 1 respectively of the noninverting inputs. The output pin 4 of each amplifier is connected to a speaker via C6, which should be 1000 μF or larger for full power capability. Component C7 is a dc decoupling capacitor and C5 is for parasitic suppression. The speakers should be 8-inch or larger.

The Q voltage for each amplifier output is approximately 4.5 volts or 1/2 of the supply voltage. The standby current for each amplifier is about 1 ampere with no signal, therefore power dissipation at no signal condition is about 4.5 watts. You can see that the power ICs will get warm. They should be on a heatsink in full operation, but they can be operated without heatsinks for short duration during experimental testing.

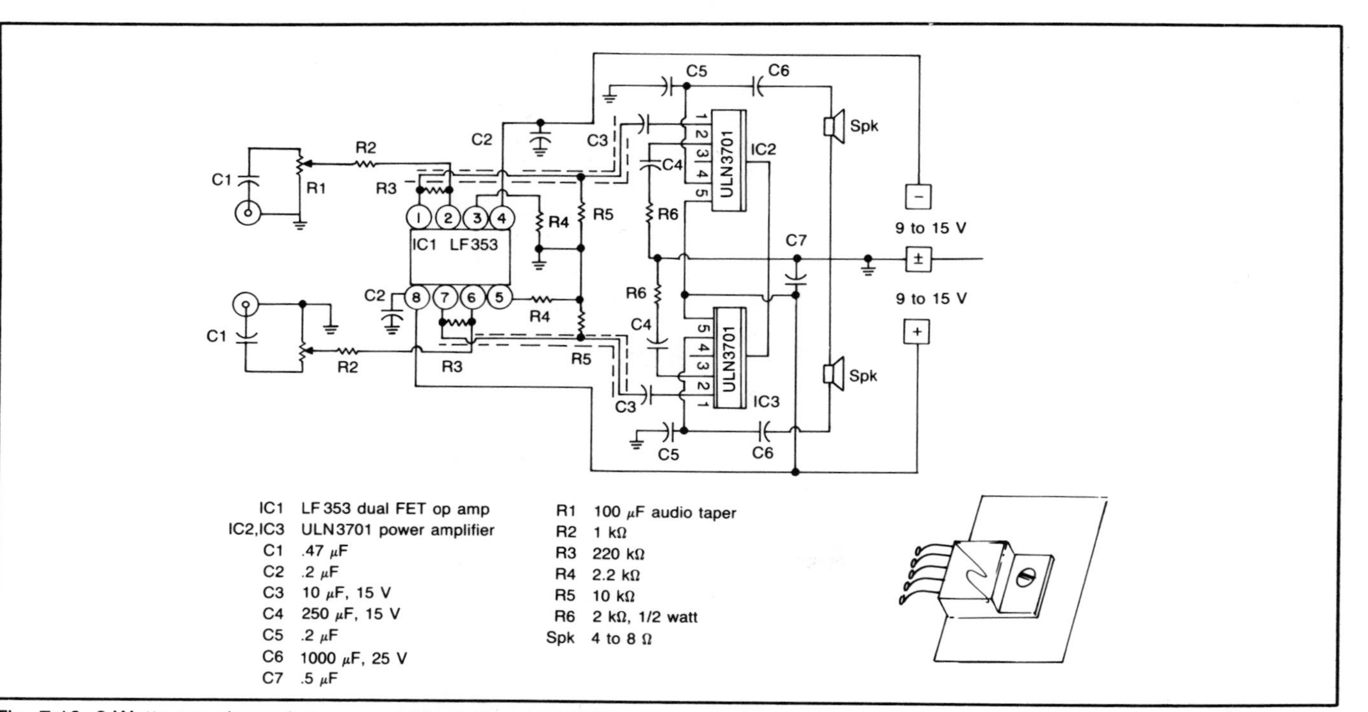

Fig. 7-16. 8-Watts per channel stereo amplifier with preamplifier.

This circuit has many outboard components for balance and compensation, which makes proto-board layout difficult without ac pickup, or noise injection by unsoldered connections. It would be better to construct it directly to an etched board with component leads as short as possible. One should utilize shielded cables on all inputs to avoid ac coupling.

Further Study

Do a frequency response plot with an audio generator and oscilloscope or ac meter. Use a common input point as mono injection of signal, and balance the outputs. Plot frequency versus amplitude of signal on graph paper from 20 Hz to 20 kHz. Keep the input level constant at each test point.

PROJECT 17
BRIDGE AMPLIFIER (12-WATT)

Function

The utilization of two LM384 power amplifiers in a bridge configuration provides heavy power capabilities with a minimum of outboard components. This circuit is ideal for a portable public address system or as a bull horn for directional communications. It requires low driving power therefore it would make an excellent phonograph or tape amplifier.

IC1 and IC2 are two LM384 power amplifiers that are connected in a bridge arrangement. A single LM384 is capable of 7.5 watts of power with an 8-ohm speaker when operating at a supply voltage of 26 V. With two LM384 ICs in a bridge design, a power capability of 12 watts is possible. The circuit shown here provided good volume from 6 to 15 volts. (See Fig. 7-17.)

The audio voltage swings the full 12 volts in a bridge design with an 8-ohm speaker as the load. It is important to balance the quiescent voltage at each output (pin 8) with R4. Alternately, place a voltmeter between minus and pin 8 of each amplifier and adjust R4 control until the quiescent voltage is the same. Capacitor C4 is necessary to balance the audio inputs to the amplifiers.

This type of amplifier must have supression of oscillation that occurs in the output during the negative swing of the voltages (R3

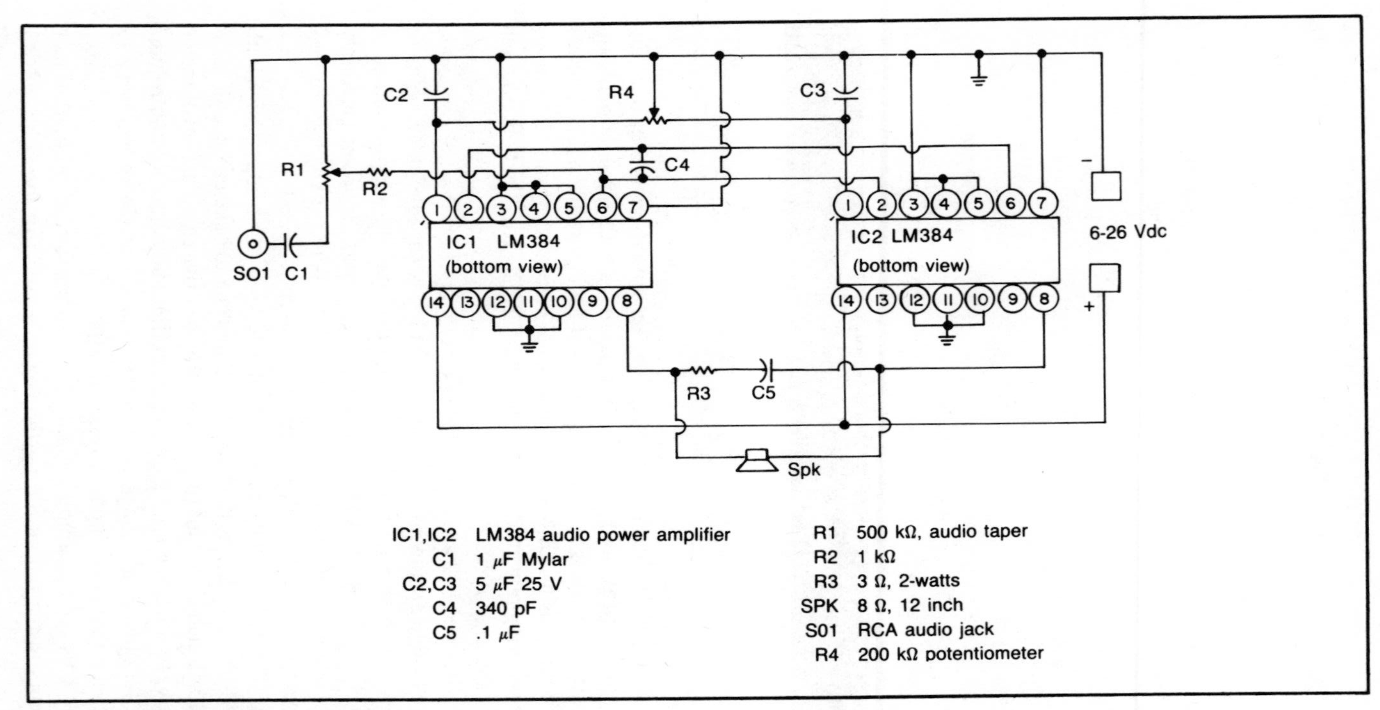

Fig. 7-17. Bridge amplifier (12-Watts).

and C5 provide the supression and must be in the circuit). Decoupling at the power inputs, pin 14, with a .1 μF capacitor to ground should be followed when the power supply is more than 3-inches from the dc inputs.

The amplifier could be driven with a minimum of 20 mV P-P of audio input. Voltage divider R1 provides for adjusting input drive to prevent overdrive distortion and should be audio taper for good control. Further input loading via R2 to the inverting and noninverting inputs of the bridge connection is provided for each amplifier.

The input from an FM radio (speaker lead connection) via shielded cable was coupled to socket SO1. The musical and voice tests revealed superb quality. An audio generator input analysis showed good response from about 70 hertz to 12 kHz. Pins 3, 4, 5, and 10, 11, 12 should be grounded to help heatsinking. No signs of overheating was observed in proto-board operation.

Further Study

Use an audio generator with input held constant and check the output response with an oscilloscope. Plot the frequency on graph paper from 70 Hz to 12 kHz. Try a tape cassette player as a drive source.

PROJECT 18
SNOOPER SCOPE (DIRECTIONAL AUDIO RECEIVER)

Function

A snooper scope provides a way of receiving and amplifying weak sounds from a distance. It is directional by way of a parabolic receiver and is useful for recording the sounds of nature such as wild birds.

IC1, LF351, is a high-performance JFET input operational amplifier with a gain of 1000 via R2/R1 (or 1 megohm/1 kilohm). The operational amplifier must have an offset voltage at pin 6 that is close to zero. Pins 1 and 5 provide offset balance via R5; set a digital multimeter between 6 and common (B+B−) and adjust R5 to read close to zero at no signal input. The output is coupled from pin 6 to R6, a signal voltage divider, for drive control at the next stage; be sure to shield this lead to avoid any ac pickup. (See Fig. 7-18 and 7-19.)

IC1 and IC2 are two LM384 power amplifiers that are connected in a bridge arrangement. A single LM384 is capable of 7.5 watts of power with an 8-ohm speaker at a supply voltage of 26 V. With two LM384 ICs in a bridge design, a power capability of 12 watts is possible. The circuit shown here provided good volume with 9 volts.

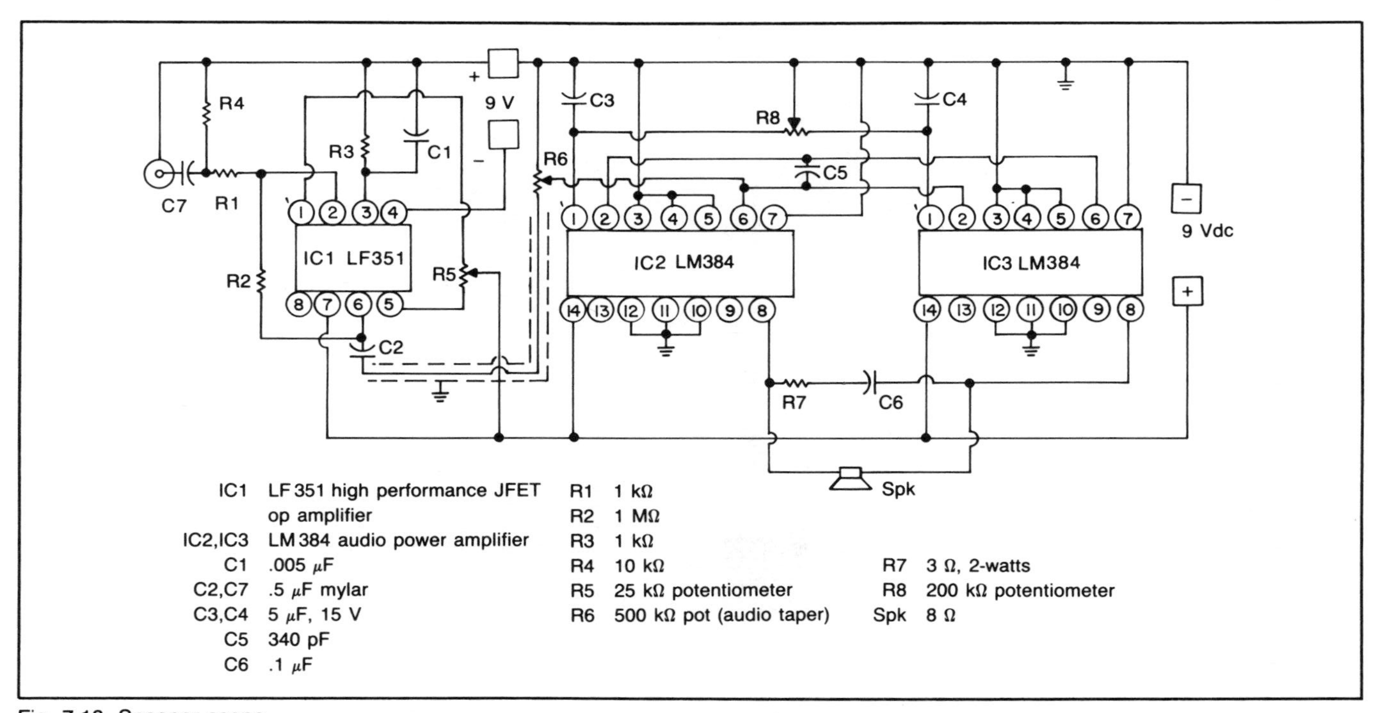

Fig. 7-18. Snooper scope.

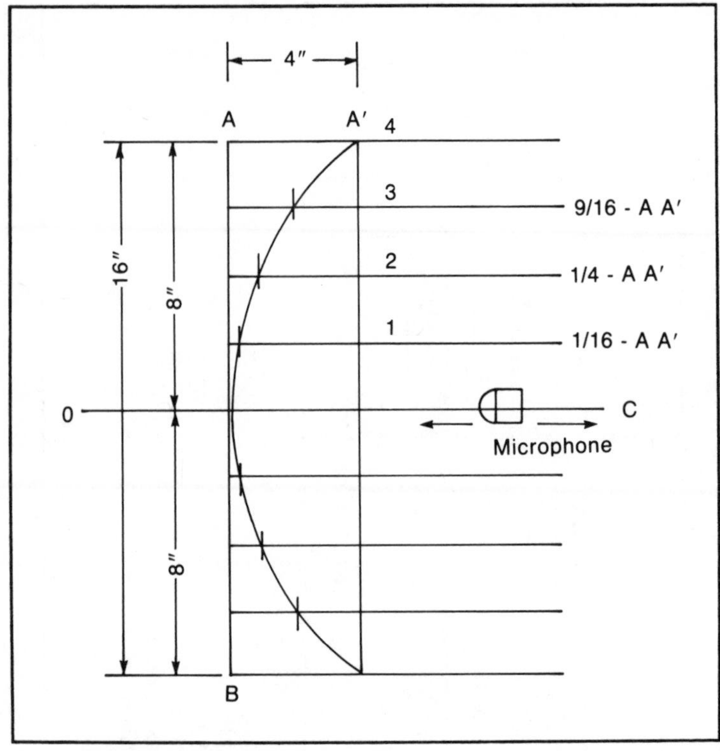

Fig. 7-19. Parabolic microphone.

The audio voltage swings the full 9 volts in a bridge design with an 8-ohm speaker as the load. It is important to balance the quiescent voltage at each output pin 8 with R8. Alternately, place a voltmeter between (common) minus and pin 8 of each amplifier and adjust the R8 control until the quiescent voltage is the same. Capacitor C3 and C4 are necessary to balance the audio inputs to the amplifier.

This type of amplifier must have supression of oscillation that occurs in the output during the negative wing of the voltages (R7 and C6 provide the supression and must be in the circuit). Decoupling of the power input voltages, (pin 14 with a .1 μF capacitor to ground) when the power supply is more than 3 inches from the dc inputs is necessary.

The amplifier could be driven with a minimum of 20 mV. P-P of audio input. Voltage divider R6 provides for adjusting the input drive to prevent overdrive distortion and should be an audio taper for good control. It should be set at low level while searching with a parabolic reflector to zero in on the audio source.

A parabolic microphone receiver can be purchased from an audio supply house or you can design your own. The microphone should be adjusted to find the apex of the sound reflection.

There should be no signs of overheating at this power level, however pins 3, 4, 5, and 10, 11, 12 should be grounded to the copper foil for heatsinking.

Further Study

Try some experimental testing like the sound from the burning of a match, the tearing of a piece of paper, or the scratching of fingernails against your palm. You may wish to add an RCA jack in the output of IC1 or the output from the bridge amplifier for recorder input.

PROJECT 19
16-WATT BRIDGE AMPLIFIER

Function

A 16-watt amplifier is possible with two amplifier integrated circuits in a bridge design. The circuit utilizes a minimum of parts yet it provides quite satisfactory results. This circuit requires a minimum drive of 100 mV which can be found from a tuner. A LM386 low-voltage audio amplifier could be added as a preamplifier.

Integrated circuits, ULN3701, power amplifiers are connected in a simplified bridge configuration. The noninverting input, pin 1, of IC1 and the inverting input, pin 2 of IC2 are connected together as the signal input. Pins 2 and 1 of IC1 and IC2 respectively are connected together via potentiometer R2 with center tap to B− source. This potentiometer must be adjusted to provide a balance in the output. Connect a digital multimeter between ground and pin 4 of each IC as R2 is adjusted until they are about equal. (See Fig. 7-20.)

Each output pin 4 is connected to the speaker. Components R3 and C2 are connected across the speaker to provide oscillation supression during the voltage swing.

The ICs tend to get warm during operation and should have heatsinks to prevent damage. The ULN3701 are in a TO 220 case in which the metal back is at ground potential like pin 3. The back should be

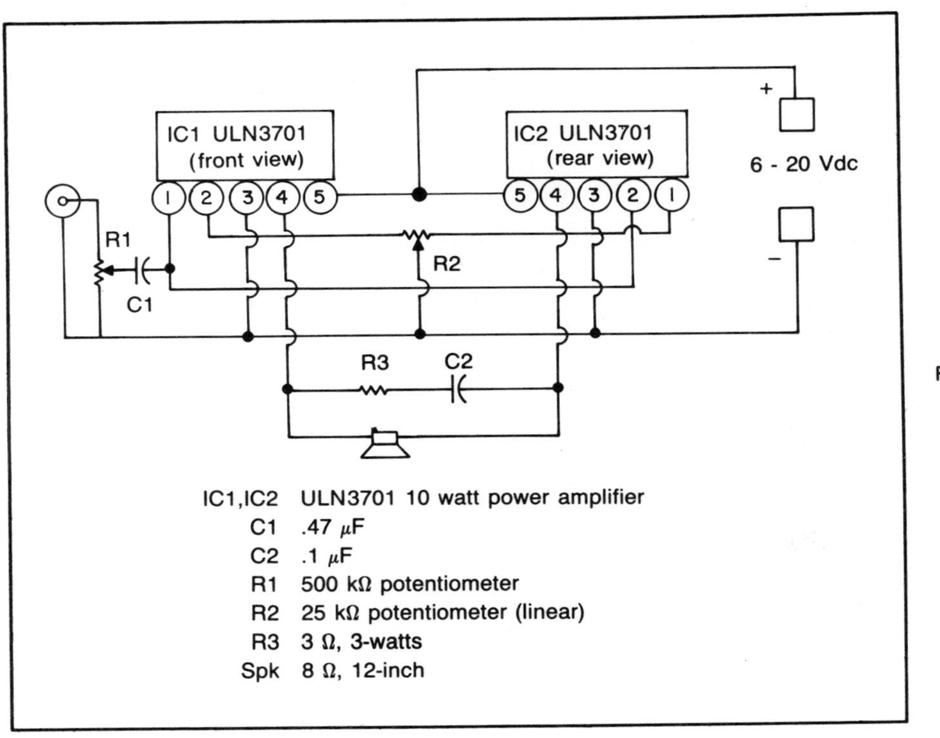

Fig. 7-20. 16-Watt bridge amplifier.

covered with a compound like silicone grease or aluminum oxide before fastening to a heatsink.

Further Study

The circuit requires considerable driving power as compared to other types of amplifiers, therefore one may wish to add a preamplifier for low-input transducers like microphones or turntable pickups.

PROJECT 20
REMOTE TELEPHONE SOUNDER

Function

In the summertime when people are out-of-doors, it is difficult to hear the telephone ring. The remote telephone sounder in this circuit sets off a tone when the telephone rings. Speaker lines can be run to the exterior of the home.

IC1 is a low-voltage audio amplifier with a pickup as the input. The pickup is placed on the telephone case next to the ringing transformer. When this is energized by the incoming signal, the 60 hertz frequency is amplified. The output at pin 5 is converted and filtered by D1, C2, R2, and C3 to produce a dc source. Dividing network R5, R4, R2 sets a threshold voltage of about .5 volts across R5, which is coupled to reset pin 4 of IC2. This voltage is caused by the offset dc level of IC1. (See Fig. 7-21.)

IC2 is an audio oscillator of approximately 500 hertz as determined by C4, C6, and R7. Pin 4 of IC2 is held at .5 volts and this is adequate to prevent oscillation.

When the pickup is brought to within 3 inches of a magnetic 60 hertz field, the voltage at enabling pin 4 of IC2 rises to approximately one volt, which is positive enough to turn on the oscillator. You will have to hunt around the case to find a close access to the ringer field.

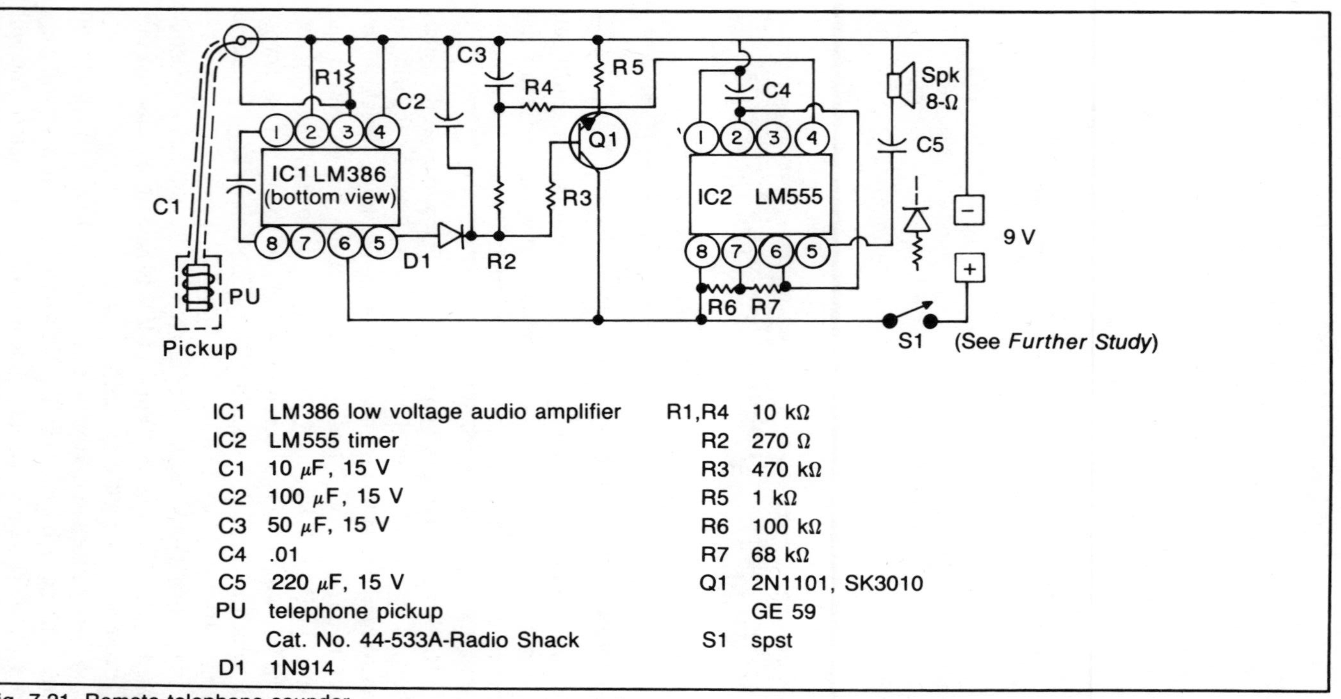

Fig. 7-21. Remote telephone sounder.

Transistor Q1 should have a beta of 50 to 70 to function properly with components shown. R3 and R5 are critical. If incorrect values are used the threshold voltage will be too high and the oscillator will be on constantly. Be sure the voltage across R5 is near .5 volts at threshold. The device can be tested with a 10 volt stepdown transformer (unloaded) as the field. Bring the pickup toward the transformer; the sound should turn on when the pickup is within three inches.

Further Study

Change the tone by changing the value of C4. Slow down the oscillator to a low pulse rate and replace the speaker and C5 with an LED and a resistor to flash a light when the phone rings. The hard-of-hearing will know the phone is ringing by observing the flashing light.

PROJECT 21
POWER LOSS ALARM

Function

A power loss alarm provides a tone to indicate such an event has taken place. This would be useful to monitor some critical concern such as a freezer or sump pump. The inoperativeness of a device might be caused by a blown fuse or a tripped breaker at the distribution box rather than utility failure.

The monitor of power occurs through a transformer, T1, and a dc (half-wave) converter D1, C1, C2, and R1 which supplies a dc voltage to IC1, NOR gate, pin 1. The truth condition for a NOR gate states: If either input is high, the output is low. One fourth of IC1, pin 2 is held low while pin 1 is held high through the dc positive voltage of the monitor. The output pin 3 is held to ground by resistor R3 and thus through connection pin 4 of IC2 is low. Since the reset pin 4 of IC2 is low, the device is disabled. (See Fig. 7-22.)

IC2, LM555, is an audio oscillator that drives a speaker. The tone is accomplished by C4, R4, and R5, which is set at 500 hertz with the components shown. When ac power is terminated, pin 1 of IC1 goes low, output pin 3 goes high and pin 4 of IC2 goes high, therefore the tone alarm is on. Standby current on the 9-volt battery is 5.3 mA while standby dc converter current from the ac conversion

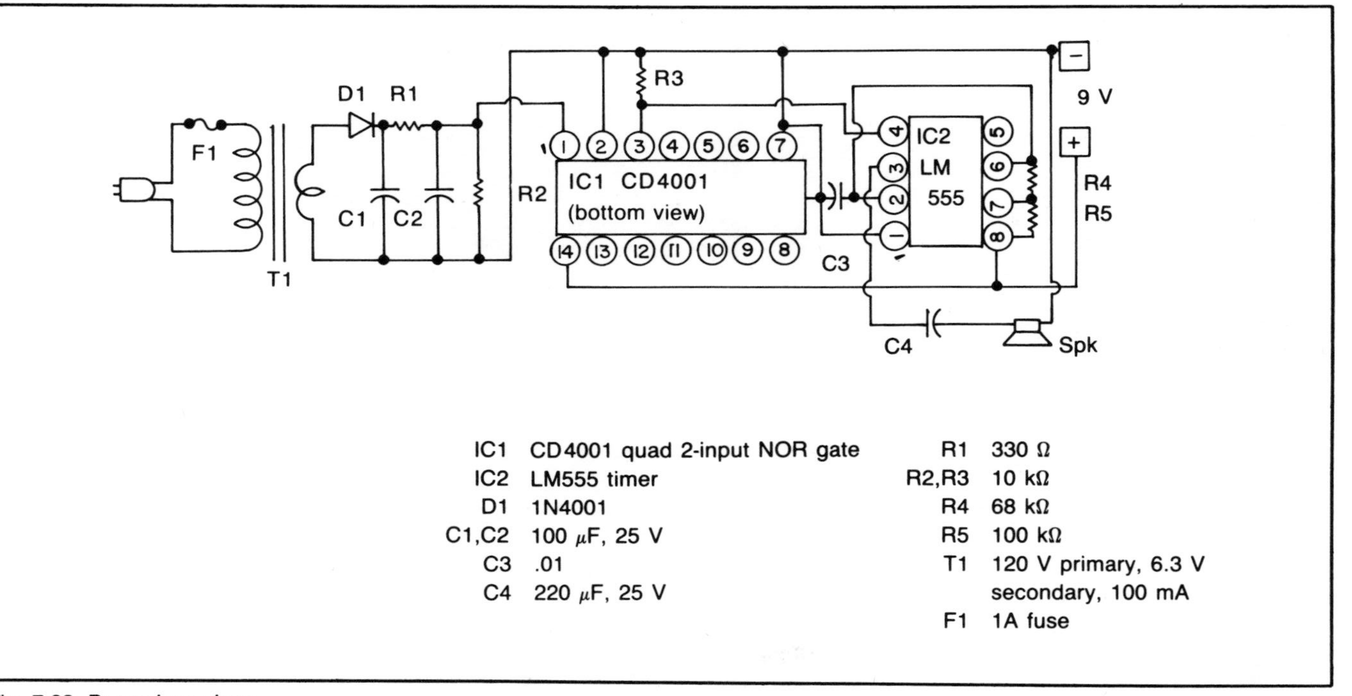

Fig. 7-22. Power loss alarm.

is 3.18 mA. The standby battery current of 5.3 mA. would pull down a small 9-volt battery to unusable state in about 24 hours. The unit operates with voltages from 4 to 12. You may wish to use a 12-volt automotive type battery for long monitoring time.

Further Study

Could you design the circuit to reverse the process? Alarm comes on to indicate power return (think OR gate). Can you think of a ni-cad battery system that is charged by the ac converter and becomes dc power for the circuit when power failure occurs.

PROJECT 22
MELODY GENERATOR/MELODY DOORBELL

Function

The melody generator in this circuit becomes a melody doorbell that sounds like chimes with two changes. The doorbell has nine available outputs for programming the tones and sequence order. Many creative combinations are possible.

The voltage-controlled oscillator section of the phased-locked loop, IC3, pin 4 is coupled to a transistor amplifier and speaker to produce an audible tone. The tone is determined by the values of C2, R9, and the voltage level at pin 9. Components C2 and R9 set the threshold level of high to low as the beginning point below which the tones will sequence. Changing C2 from .01 to .03 will start the tones lower and they will sequence lower. (See Figs. 7-23 and 7-24.)

IC2, MC14017, is utilized as a decoder which shifts (switches) one position with each pulse of the clock. Although the position shift is in logical order, one can set the order of how the sequence is to take place. (See explanation in next paragraph.) Pin 4 is the input for the clock, pin 15 is reset, and pin 13 is clock enable. The circuit is operated as a melody generator (free running) when both 15 and 13 pins are held to minus. It can be made to cycle through and halt (see dash lines on schematic for change) by coupling pin 13 enable

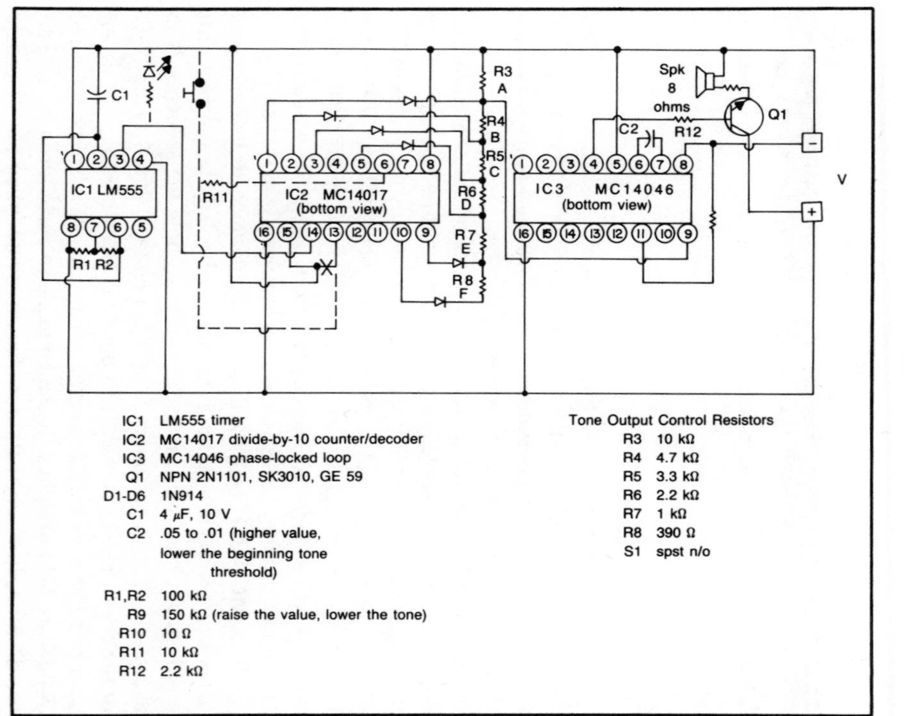

Fig. 7-23. Melody generator/melody doorbell.

to the last open pin in the sequence, this will pull pin 13 high at the end of the cycle and thus halt. Switch S1 triggers the enable to low and thus starts the cycle over.

The tone is set by a voltage divider network R3 through R8. The VC oscillator, IC3 pin 9, has the voltage set at point A. When pin 1 of IC2 goes positive through the diode, approximately 8 volts appear at that point and the tone output is the highest, but pin 1 is the 5th order in sequence of shift. The 4th order of sequence is pin 10 and since it is connected to the last resistor in the divider, its tone control is lower, thus the 8 volts that was at point A is now at F. The dividing network produces 3.5 volts at point A or the lowest tone in the sequence.

If one tailors carefully the resistors R3 through R8, a number of tones can be planned. Only six tones were planned in the schematic, nine tones could have been utilized with additional diodes and resistors. The decoder is a divide by ten IC, but we needed the last pin to shut-

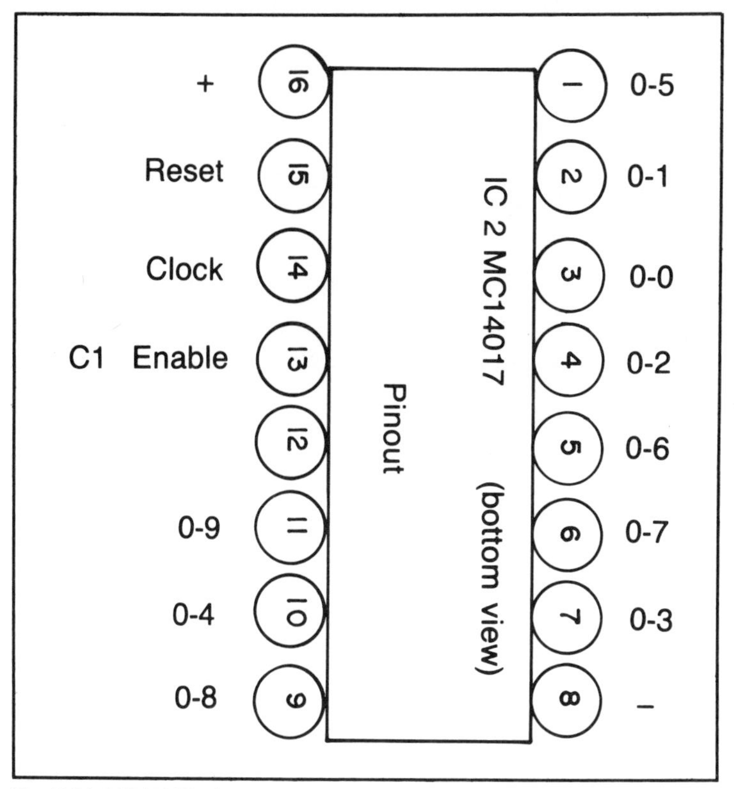

Fig. 7-24. MC14017 pinouts.

down the cycle, thus only nine notes are possible. One can shift the diode's positions also to determine the order of the outputs. (See Fig. 7-24.)

It is a good practice to set values of R3 through R8 to tone choices. This can be done for the doorbell chimes with the circuit intact. The procedure is to connect a jumper wire to the positive and then probe points A, B, C, D, E, and F one at a time. Listen to the tone at each point and if you wish to raise or lower the tone slightly, change one of the resistors.

IC1, LM555 timer, is a clock that provides the sequence transition of IC2. It is designed to provide one pulse per second. Increasing or decreasing the value of C1 will change the tone delivery rate.

Further Study

There are many suggestions in the text above that will help one to learn what happens when changes are made.

PROJECT 23
ELECTRONIC CALLER

Function

An electronic caller would be useful to call the family to dinner, to call a class together for some event or announcement, or to call a meeting together with an electronic gavel.

IC2, LM386, is an oscillator-amplifier with C2, R2, C3, R3, and R4 to determine the frequency and waveshape. Component R4 can be adjusted to provide a frequency output from 50 to 1500 hertz. The frequency shifts upward on turn on, holds steady, and shifts slightly down at shutdown. The output is coupled through C4 and R6 to Q1, which drives the speaker. (See Fig. 7-25.)

The power transistor Q1 may be omitted to save on parts cost by connecting C4 directly to the speaker as shown in the schematic in dashed lines. It is possible to connect the speaker directly to pin 5, however, it changes the loading and thus causes a change in frequency that can be adjusted to your choice. You can experiment with any of these options.

IC1, CD4011 NAND Gate, is a time ON delay to OFF control. The truth logic states that when two gates are high, the output is low. Under standby state, gate 2 pin 10, is low because pins 8 and 9 are high. The output of gate 1 pin 3, is high and connects directly to pins 8

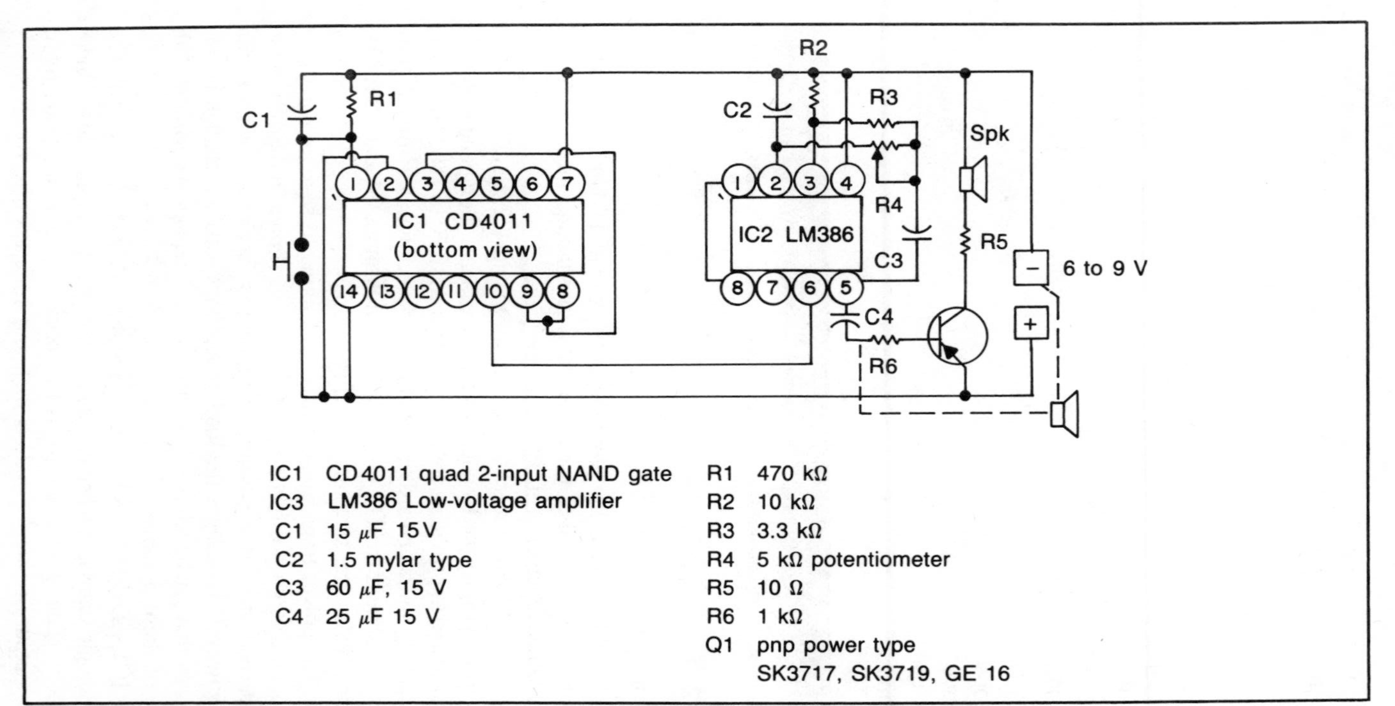

Fig. 7-25. Electronic caller.

and 9. Gate 1 pin 2 is connected to positive or high, while pin 1 is connected to ground through a 470 kΩ resistor. Capacitor C1 is connected to ground from the same pin and connected also to switch S1. When S1 is closed momentarily, C1 charges instantly to full voltage and all the outputs turn ON in IC1. When S1 is released, C1 discharges through R1 at a time constant of about five seconds, thus the sound is ON for that length of time. Increasing the value of C1 or R1 will change the length of time ON.

Further Study

Use two digital voltmeters to monitor the changes in the gates to observe the logic of the truth statement. Place one digital voltmeter between pin 1 and ground to observe the action of pin 1 as capacitor C1 charges and discharges. Place the other digital voltmeter at pin 6 and ground to observe the output condition.

PROJECT 24
TREMOLO

Function

The tremolo is a sound-effect device that alters the signal of a string instrument. The voltage-controlled oscillator of IC1 (phase-locked loop) produces low square-wave frequency from 5 hertz to 2500 hertz that is controlled by varying the voltage at pin 9 with R2. The frequency threshold is set by R1 and C1. Increasing the value of R1 will increase the frequency, whereas decreasing the value of C1 will decrease the threshold frequency. (See Fig. 7-26.)

Q1 is operated as an amplifier with a Q point of four volts with a transistor of 100 beta. The gain without tremolo effect is about two volts. With bypass capacitor C4 in the circuit the gain is 33. Instrument drive cannot be greater than 30 mV with C4 in the circuit (no tremolo) to avoid distortion. The drive can be 1.5 volts maximum with no distortion with C4 out of the circuit on a straight-through mode.

The output from IC1 is coupled to the collector of Q1 via depth control R4. The square-wave signal pulls the audio at a frequency determined by R2. The oscillator is activated by foot switch S1.

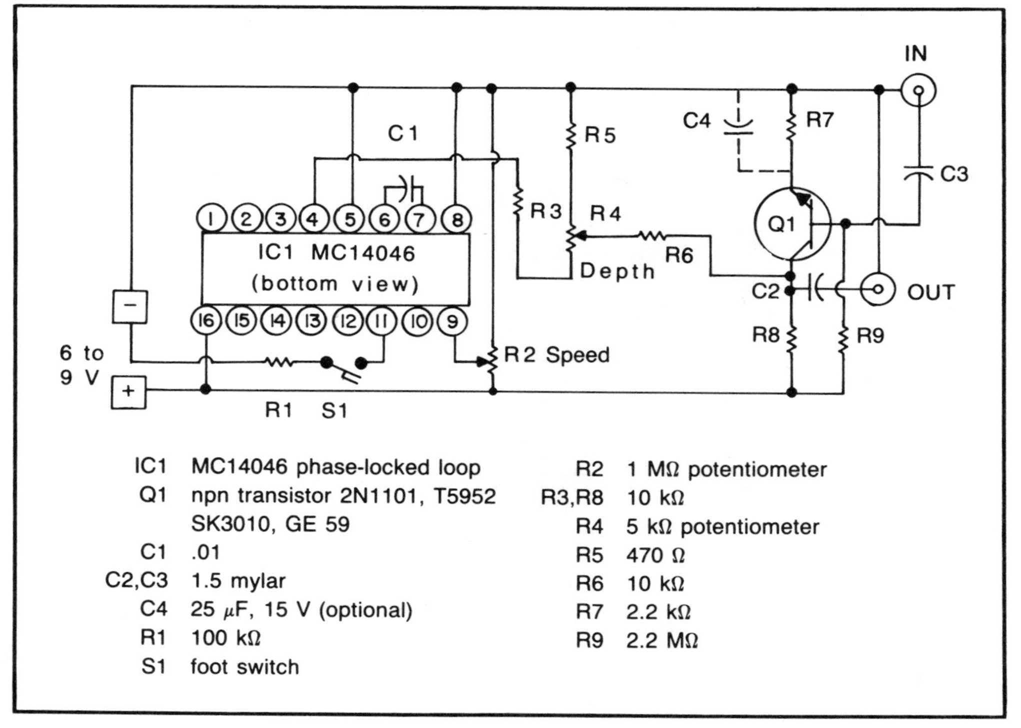

Fig. 7-26. Tremolo.

Further Study

Change values of C1 and R1 to note the effect on the square-wave threshold. Vary speed and drive to note frequency change and amplitude. Try C4 both in and out of the circuit.

PROJECT 25
SOUND/VIBRATION ALARM

Function

The sound/vibration alarm is an excellent monitor of a state or condition. It could be utilized as a deterent to an intruder or the warning of excessive vibration of a piece of machinery.

IC2, UA556 timer, pin side 8 through 11 is a one shot multivibrator which is triggered at pin 1. The length of time ON is determined by C7 and R9. Increasing the value of either will increase the ON time. The LED in the output pin 9 monitors the ON state. (See Fig. 7-27.)

Pin side 1 through 6 is an oscillator with C6 as the major controller of tone. Increasing the value of C6 will increase the tone. In fact, a high pitch tone is better to scare an intruder. The reset pin 4 is connected to the output pin 9 of the one shot and to ground through R6. The reset turns on the oscillator when pin 9 goes high.

IC1, LM386, is connected as an amplifier with a gain of 200. The output pin 5 is rectified by D1 and R5. Component C4 charges through R11 when first turned on. When the sound appears in the output and is converted by D1, C4 discharges providing a spike at pin 8 of IC2, thus the one shot is triggered ON and the tone output from the oscillator is ON.

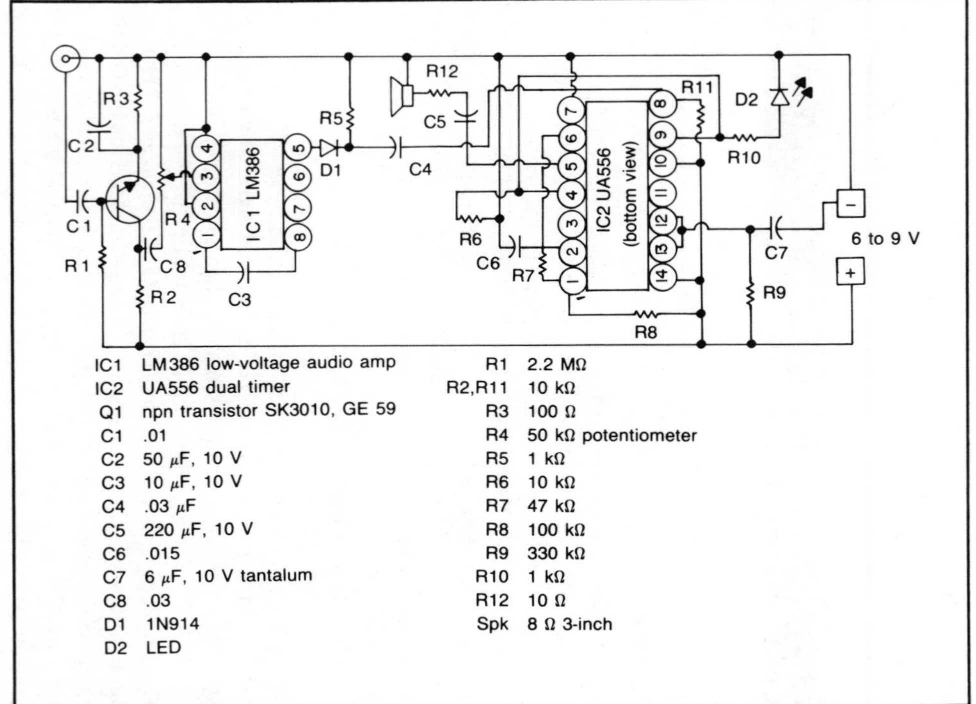

IC1	LM386 low-voltage audio amp	R1	2.2 MΩ
IC2	UA556 dual timer	R2,R11	10 kΩ
Q1	npn transistor SK3010, GE 59	R3	100 Ω
C1	.01	R4	50 kΩ potentiometer
C2	50 μF, 10 V	R5	1 kΩ
C3	10 μF, 10 V	R6	10 kΩ
C4	.03 μF	R7	47 kΩ
C5	220 μF, 10 V	R8	100 kΩ
C6	.015	R9	330 kΩ
C7	6 μF, 10 V tantalum	R10	1 kΩ
C8	.03	R12	10 Ω
D1	1N914	Spk	8 Ω 3-inch
D2	LED		

Fig. 7-27. Sound/vibration alarm.

Transistor Q1 is a preamplifier with a beta of about 100. A transistor with a higher beta would increase the sensitivity, however, one should change the value of R1 to set the Q point at 4.5 volts or about half the supply voltage. The output at the collector is coupled through C8 and R4 to ground. Component R4 provides a method of adjusting sensitivity. When full on with the transistor listed, the alarm will go off with a clap of your hands at about four feet from the mike. If you make a cone receiver in which to place the mike, the alarm will go off at about eight feet.

The vibrator monitor is accomplished by taping the mike to a surface. A tap with a pencil set the alarm off. Remember to change the one shot time if you want the alarm to be ON for a longer period of time.

Further Study

Change C7 or R9 to increase the ON time. Try changing C6 to a different tone. Adjust R4 to determine levels of sensitivity. Make a cone concentrator to increase the sensitivity and sound direction.

PROJECT 26
TUNABLE TONE GENERATOR/AMPLIFIER

Function

A tunable audio generator is useful for sound affects or as a test generator for audio amplifier testing.

The voltage-controlled oscillator of the phase-locked loop, IC1, provides an adjustable tone from 30 Hz to 10 kHz by varying the voltage at pin 9 through voltage-divider potentiometer R2. Changing resistor R1 will also affect the frequency output range. Lowering it to 100 kΩ will produce a tunable range from 166 Hz to 25 kHz. Capacitor C2 affects the output frequency. Increasing it to .003 μF will lower the frequency. (See Fig. 7-28.)

The output of IC1, pin 4 is coupled through C1 and R3 to negative. The potentiometer divides the output and sets the level to input pin 6 of IC2, a dual 2-watt audio amplifier. R3 should be an audio taper to control the volume less harshly. The output pin 2 of this stage is connected to C4 and on through to an 8-ohm speaker.

IC2 is a dual 2-watt amplifier (one half used) that drives a 12-inch speaker with no difficulty, so go easy on the volume control. A feedback circuit R4 and C3 is connected from pin 7 to negative. Without this the IC produces self-oscillation in the output, which reduces the quality of tone. The manufacturer recommends a 2 kΩ resistor, how-

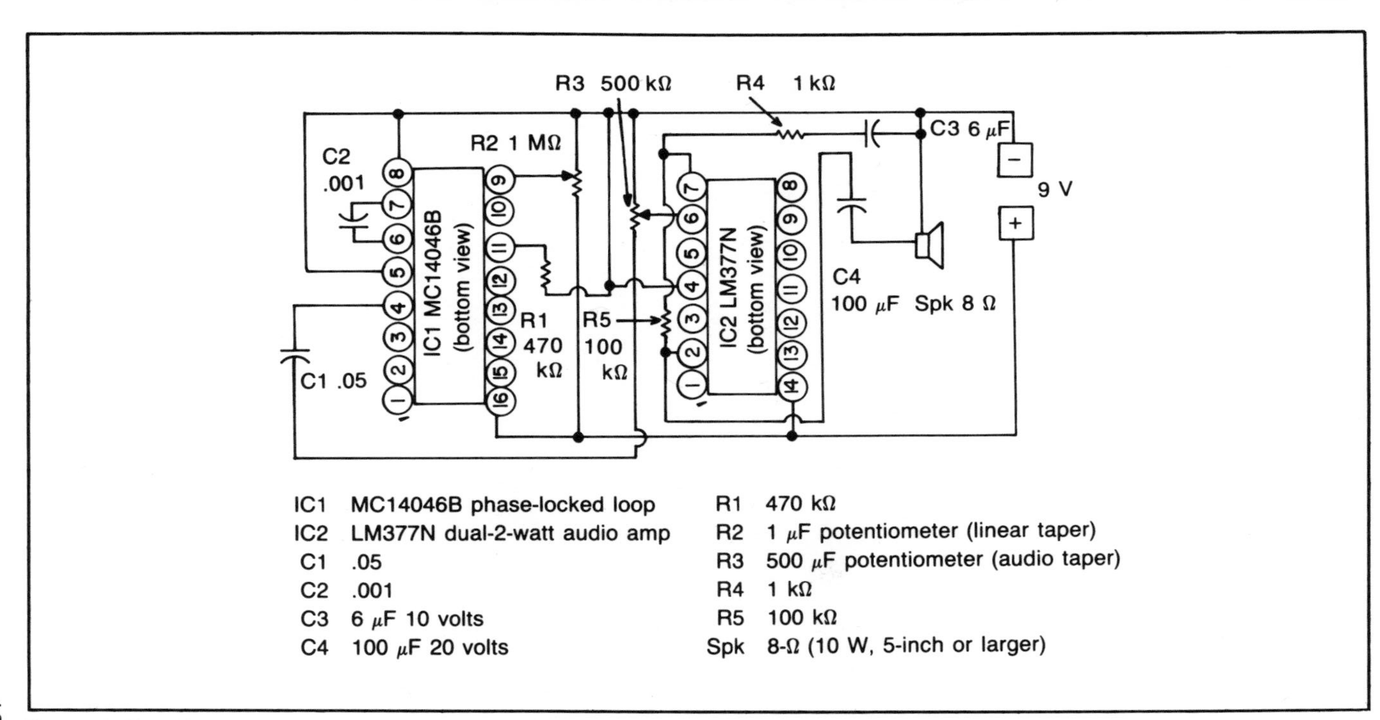

Fig. 7-28. Tunable tone generator/amplifier.

ever a 4.7 kΩ removed the oscillation completely. The output is very close to a square wave, and thus it makes a good audio generator for testing purposes.

Further Study

Change the value of C2 and note the change in the base level of the frequency. Add a spdt switch, RCA phone jack, and a 10-ohm resistor at the input to the speaker to make the unit into an audio tester.

PROJECT 27
FLUID DETECTOR ALARM

Function

A fluid detector with ac or dc output has many useful applications. One application would be in the bilge of a boat to warn of an inoperative sump pump or leaks greater than the pumping capacity. It could be used any place where rising water represents a threat. The IC, ULN2429A is unique in that the probe sensors are in ac operation when activated by liquid to prevent plating problems at the probes. (See Fig. 7-29.)

IC1 has an output at pins 1 or 14 at a frequency of 400 hertz as determined by C2. When a conductive liquid touches the contacts of the probes, the internal oscillator signal is coupled to the internal detector and a square wave appears in the output. The manufacturer recommends both outputs, pins 1 and 14 be utilized when heavy current load is required. The IC has a maximum rating of 700 mA in the outputs but in this circuit design, only 14 mA was used to light the LED. The output drives the LED and is coupled to IC2, pin 3 via C3 and R3. Component R3 sets the drive level and thus the loudness.

IC2 is a low-voltage audio amplifier that was explained in previous applications. Component C5 is optional and could be added where

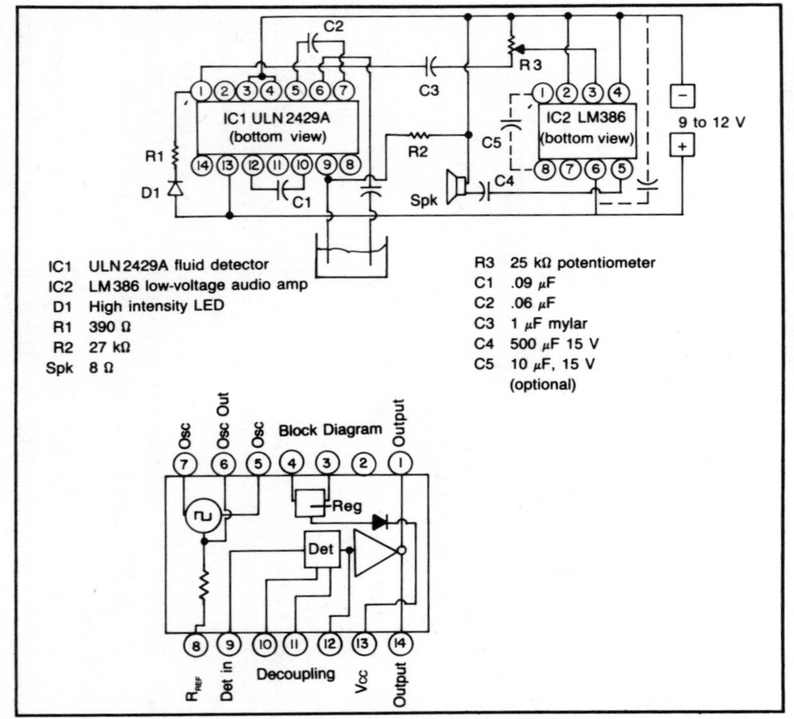

Fig. 7-29. Fluid detector alarm.

greater amplification is desired. Capacitor C1 is recommended for decoupling between pins 10 and 11 of IC1, however the tone quality was improved by decoupling between pin 10 and 12.

Further Study

One can omit the audio amplifier IC2 and change C2 to 10 μF to produce a flashing LED when probes become emersed (this is a silent sentry). Change the value of C2 and note the change in the tone. Scope the output waveform and note waveshape and time.

PROJECT 28
TWO-STATE MONITOR/SOUND ALARM

Function

The Two State Monitor/Alarm accomplishes two goals: it scares off an intruder by the element of surprise and a loud sound, and it notifies security that an intrusion has taken place. (See Fig. 7-30.)

IC1, CD4001, is a two-input quad NOR gate. The truth table of this gate states when both inputs are LOW, the output goes HIGH. The inputs of two gates, pins 2 and 6, are tied to B− or held LOW. The other two inputs pins 1 and 5, are held HIGH through inputs 1 and 2. They are held also to B− via R2 and R1. The output pins 4 and 3 are connected to IC2, pin 4 enabling via R5 and R6. Also monitor LEDs, D2 and D1, are connected to B− via R3 and R4.

If an intruder opens input 1 or 2, the NOR gate makes both inputs LOW and the output goes HIGH. The LED comes on to monitor the intrusion location 1, 2, or both.

IC2 is an oscillator that is turned on because pin 4 goes HIGH as a result of the gate action. The sound tone is controlled via C1, R7, and R8. The output pin 3, is coupled to an 8-ohm speaker via C2. Capacitor C2 should be 300 μF or larger to improve efficiency and increase loudness.

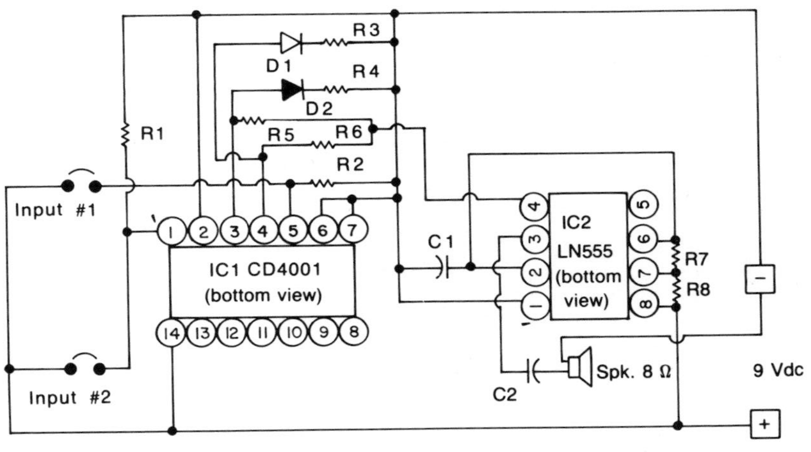

IC1	CD4001	R1,R2	10 kΩ
IC2	LN555 timer	R3,R4	470 Ω
C1	.006	R5,R6	1 kΩ
C2	300 μF, 15 V	Spk	8 Ω
D1,D2	red LEDs		

Fig. 7-30. Two state monitor/sound alarm.

Further Study

Change the value of C1 to program tone (lower value will decrease tone, higher value will increase tone). Use a 15-inch speaker and then a 4-inch speaker to note efficiency and loudness. Try lead foil as a jumper and then melt it with a flame to simulate a fire warning.

PROJECT 29
FILM PROCESSING TIMER (LIGHT/SOUND)

Function

A film processing timer to communicate the completion of film developing is useful for photography. Monochrome film such as Pan X or Tri X requires specific time in the developer at 68 degree solution temperature when developing with Microdol Developer. Fixing time of 10 minutes is recommended also. Whether in the darkroom or in light by tank process, keeping track of time requires concentration. The circuit here is a preset timer of 7 and 14 minutes that communicates the end of these periods by light or sound, thus one can devote his work to other activities and yet be reminded. (See Fig. 7-31.)

CD4081 is a quad two-input AND gate in which two gates are utilized as a timer circuit. The truth table of an AND gate states that when any two inputs go HIGH, the output goes HIGH. The two gates are cascaded in order that the timing of gate 1 (when turned on) starts the timing of gate 2 to provide a cumulative time span.

Only one side of CD4081 is utilized in the time design. Pin 1 of gate 1 and pin 6 of gate 2 are held high to the B+. Pin 2 of gate 1 is in a (t=RC) time constant via R2, R1, S1, and C1. If S1 switches

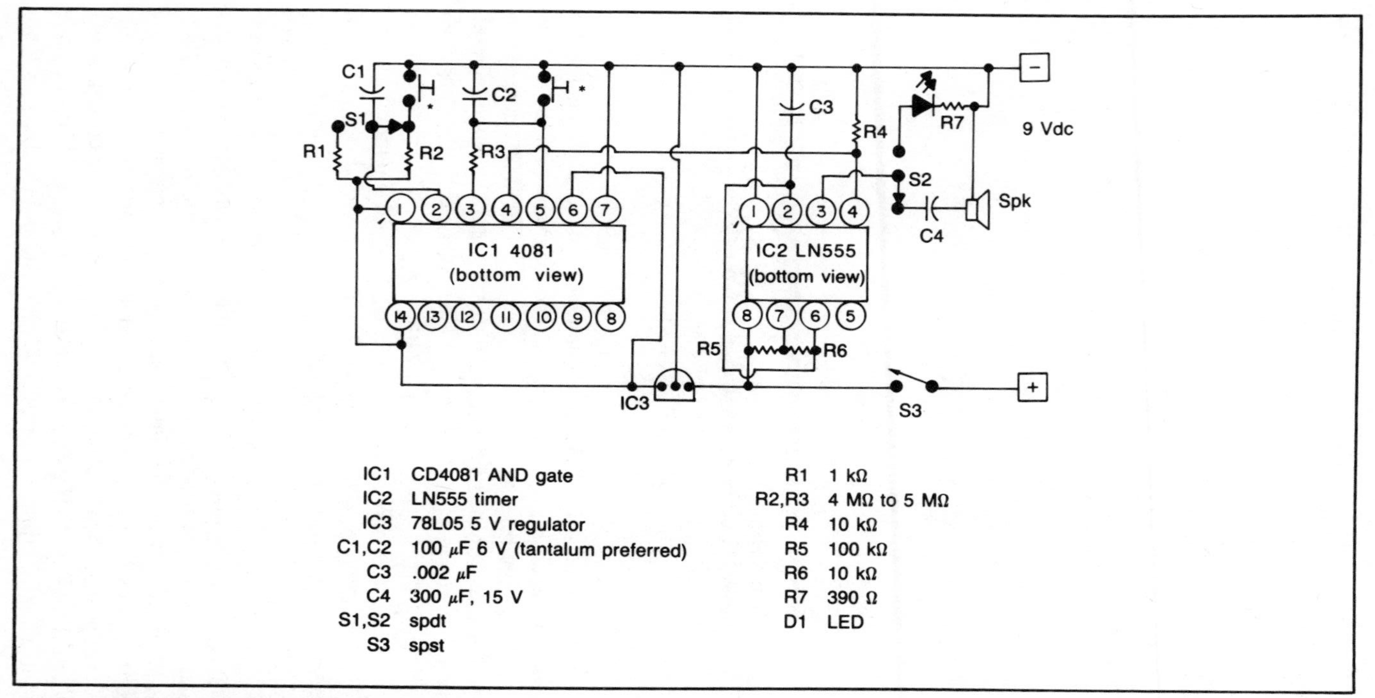

Fig. 7-31. Film processing timer (light/sound).

to R1, the time constant is short and output pin 3, goes HIGH immediately. The gate 2 pin 5 is tied between C2 and R3, which is the time constant of gate 2. Timing of gate 2 is about 7 minutes.

If S1 in gate 1 is to the right, the time for pin 2 to go HIGH is longer (about 7 minutes). At the end of the 7 minutes, pin 2 is HIGH (about 2.5 volts with regulated supply of 5 volts); output pin 3 goes HIGH and starts the time charge for C2 and R3. In about 7 minutes, pin 5 goes HIGH (again 2.5 volts) and output pin 4 goes HIGH.

In the interim period, enabling pin 4 of IC2 (the audio oscillator) is held to ground via R4 and the oscillator is off. However pin 4 of IC2 is also connected to pin 4 of gate 2, thus it went HIGH to enable the oscillator to produce a tone from the speaker or turn on the LED depending on the position of switch S2. The red LED is the silent readout and might have optional use in the darkroom.

IC3, 78L05, is in the circuit to produce a constant voltage for the timer stages. This is to insure accuracy by compensating for 9-volt battery wear.

One needs to make some adjustments of C1, C2, R2, and R3 to bring the time to exact minutes. Stock capacitors and resistors were utilized and variance in tolerances required adjustments to get precise times. For example, 100 μF capacitors were 110 and 108 μF; resistors marked 4 megohms were 4.4 and 4.2 megohms. The oversizes in this case pulled the time to the higher side. Note that the high level for change was 2.5 volts with a 5-V supply or 50% of the applied voltage rather than the (t=RC) time constant of 63% of the applied voltage.

Further Study

Change the tone of the oscillator by changing the value of C3. Also change the value of R1 to provide an overall time change of eight minutes before the tone is heard.

PROJECT 30
TONE MONITOR AND/OR DETERRENT

Function

A tone monitor can act as a locator of intrusion as well as a deterrent by the element of surprise when the alarm is tripped.

IC1, CD4071, is a quad two-input OR gate with a truth table that states that when either input goes HIGH the output goes HIGH. Two gates, 1 and 2, are utilized for a two location monitor. In gate 1 pin 2 is held LOW to the B− and control pin 1 is held LOW via an input switch, which can be a fine wire. If an intruder trips the switch and opens the circuit, pin 1 goes HIGH and the output pin 3 goes HIGH via R1. Since enable pin 4 of IC2 (tone oscillator) is held low via R3, the oscillator is off, however pin 4 is connected to output pin 3 of IC1 and when pin 3 went HIGH the oscillator turned on to provide a tone via the speaker. The tone is distinct for input 1 and one can identify the location. (See Fig. 7-32.)

Gate 2 of IC1 is held LOW via pin 12 to B− and input 2 to B− via a switch. If an intruder trips the switch at pin 13, output pin 11 goes HIGH and since it is connected to enable pin 4 of IC3, the tone oscillator comes on. The frequency of this oscillator via C2 is a higher pitch than the IC2 oscillator because C2 is a smaller capacitor. The differentiation of tones identify which input was tripped.

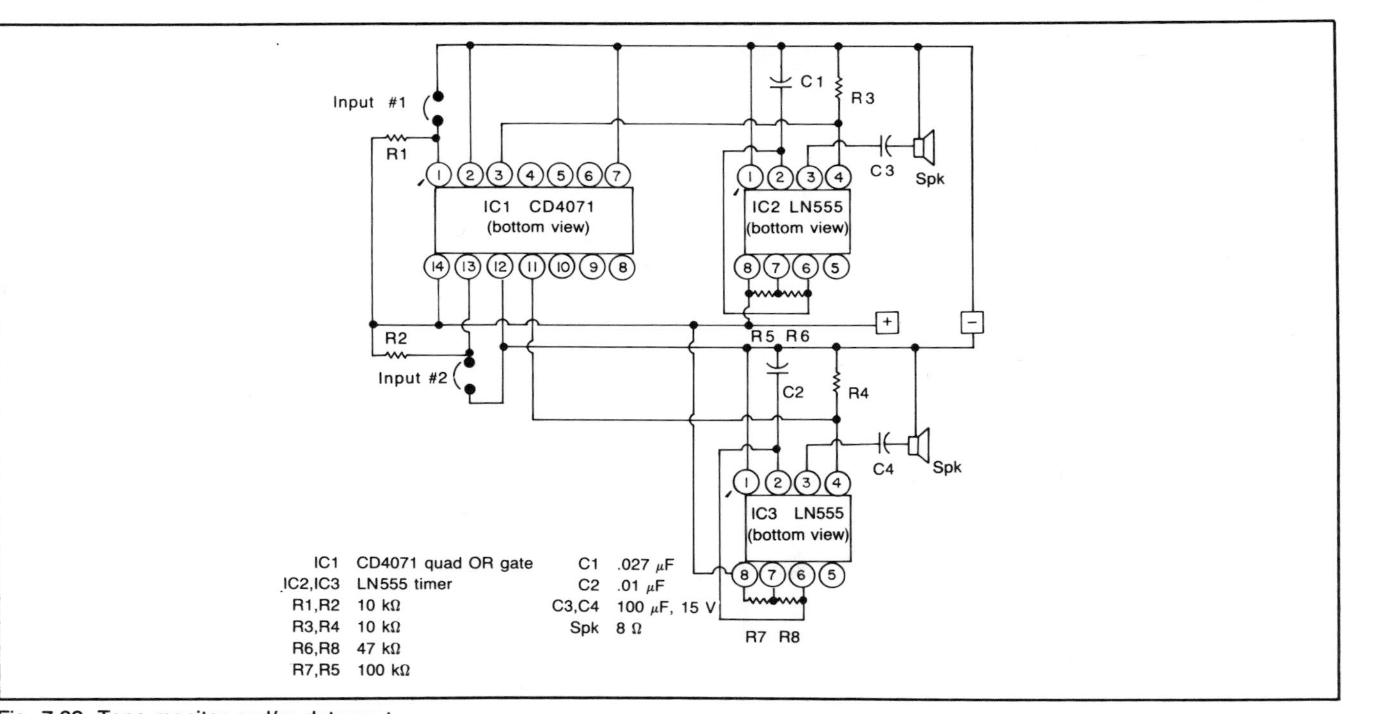

Fig. 7-32. Tone monitor and/or deterrent.

If both switches were opened in sequence, one would hear a low tone, a higher tone, and then a harmony of both tones. In this manner, one can identify either or both intrusions by the same intruder moving about or a different intruder. In addition, the multiple tones surprise the intruder and hastens departure.

The LN555 is more than adequate to drive a large speaker. If one desires a louder sound, you can accomplish this by coupling the output to a powerful amplifier, or you could use one of the amplifier ICs used elsewhere in this test.

Further Study

Can you add a third and fourth input to produce additional monitors and tone oscillators? Can you change C1 or C2 to produce other tones?

PROJECT 31
BACKUP BEEPER (VEHICLE BACKUP WARNING)

Function

Drivers of trucks and other heavy vehicles have difficulty in seeing directly behind when backing. A backup beeper warns pedestrians that a vehicle is about to proceed backward. This circuit simulates a backup warning by producing a beeping tone at the rate of about 1 per second. The loudness is adequate without additional IC amplification although one could be added. (See Fig. 7-33.)

IC1 is a dual timer in which side 1 is a slow pulse oscillator of about 1 hertz per second. Components R2, R1, and C1 are the long time constant. Time on can be calculated by $t_1 = .7(R1+R2)C1$ or $t_1 = .7\ (100\ k\Omega+220\ k\Omega) \times .000005$ or 1.15 seconds. The off time is shorter than the on time; $t_2 = .7\ (R2)\ C$ or $t_2 = .7\ (200\ k\Omega) \times .000005$ or .77 seconds. The total time for 1 Hz is 1.92 seconds. Enabling pin 4 (reset) is held HIGH to keep the oscillator free running when voltage is applied to pin 14. The output pin 5 is coupled to enabling pin 10 of side 2.

Side 2 is an audio oscillator of about 1000 hertz as determined by C2, R3, and R4. Reset or enabling pin 10 is connected to output pin 5 of side 1. When output pin 5 switches to the positive at 1 times per second, the oscillator beeps a short pulse tone of 1 kHz.

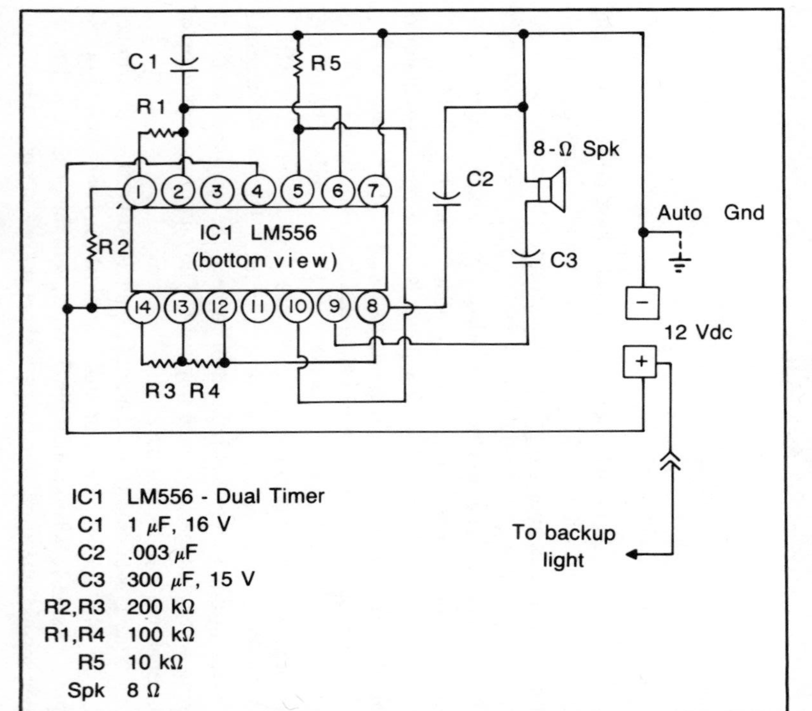

Fig. 7-33. Backup beeper (vehicle backup warning).

Coupling to a back-up vehicle would be relatively easy since 12-volt power potential is available when the backup lights are turned on and when the vehicle is shifted in reverse. The B+ terminal on the circuit would need to be spliced to the B+ backup light lead in the vehicle. A 12-volt power supply can be utilized for laboratory testing and analysis.

Further Study

Measure output time on an oscilloscope at pin 9. This would be easier to read if side 2 was free running by disconnecting the lead going from pin 10 to 5 and connecting it temporarily to B+ to enable.

Measure the pulse frequency with an oscilloscope on side 1 pin 5. Again, this can be done with the lead going from pin 5 to pin 10 disconnected, thus eliminating the bothersome tone during testing. You cannot leave pin 10 floating so be sure to connect a jumper temporarily from pin 10 to negative to insure shutdown.

PROJECT 32
TOXIC GAS ALARM

Function

The toxic gas alarm warns of a dangerous situation such as the presence of an explosive gas or contaminated air that is hazardous to one's health. Vapors or gases from such items as alcohol, lacquer thinner, benzene, propane, butane, carbon monoxide and other hydrocarbons can be detected by the TGS 812 sensor according to the manufacturer. (See Fig. 7-34.)

IC1, LM555, is utilized as an oscillator that produces a high pitched sound of 1600 hertz that is similar to a smoke alarm. The oscillator is enabled by pin 4 (reset); when this pin is high, the sound is heard and it is off when the pin is low.

IC2, CD4071, is a quad two-input OR gate with a truth table that states that when either or both pins go HIGH the output goes HIGH. One fourth of IC2 is the control gate in which pin 2 is held low; pin 1 is held low via R3 and S1. Output pin 3 is low and it is connected to pin 4 (reset) of IC1, therefore the oscillator is off. Pin 1 of IC2 is connected also to the B+ via SCR and D1: therefore if the SCR is turned ON, IC2 switches to enable IC1 via pin 4 (reset).

The gate of the SCR is tied to B− via R4, which is in series with the *sensor* to the B+. Resistor R4 is adjustable to control the sen-

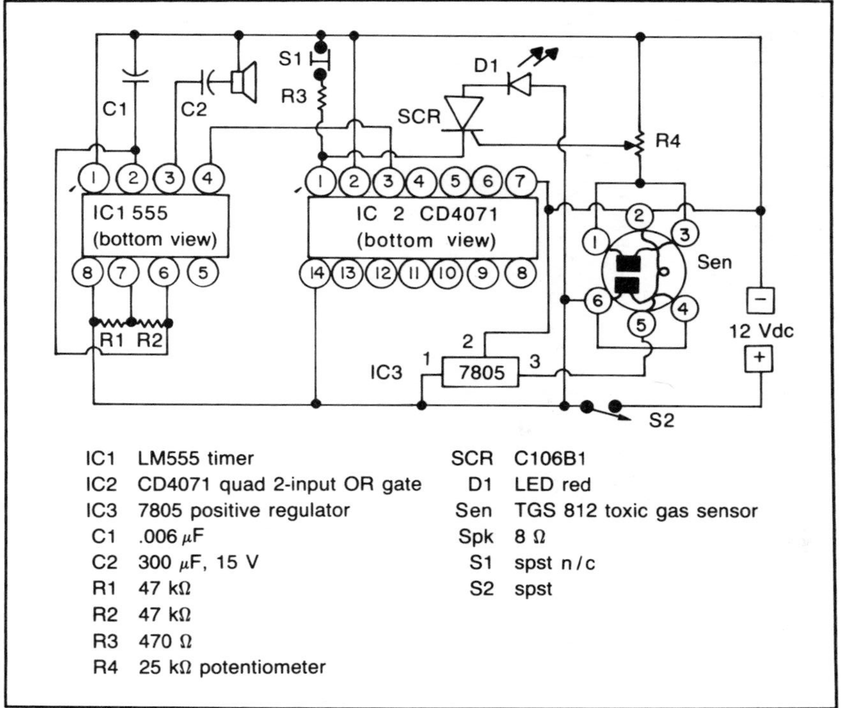

Fig. 7-34. Toxic gas alarm.

sitivity of the gate of the SCR. When the sensor conducts, the voltage across the R4 voltage divider turns on the SCR to make pin 1 of IC2 go HIGH which in turn sets the output pin 3 and enable pin of IC1 to set off the alarm tone. The LED is in the circuit to be a visual indicator in case of IC1 failure and also to monitor the switching of the SCR.

The gas sensor TGS 812 sets off the alarm in the presence of vapors or gases. When toxic gases come in contact with the sensor, a decrease in resistance causes current to flow through the load, R4. The voltage developed across the centertap and ground side of the potentiometer gates the SCR. The SCR latches when activated thus the alarm is continuous until deactivation switch S1 is pressed. The switch must be held for 12 seconds after removal of the vapor to allow for restabilization of the sensor. The manufacturer's guidelines call for 1 to 2 minutes recovery time but this was not necessary in our tests.

IC3, the 7805 regulator provides the 5 volts for the sensor filaments. Current draw in operational mode is 211 mA. Standby current is 135 mA of which 120 mA is heater current. This kind of drain would require an automotive battery supply or 120 Vac converter.

Further Study

Try utilizing the second gate as a buffer by connecting pin 3 of IC2 to pins 5 and 6, and then pin 4 to pin 4 of IC1. Try different vapors to the sensor; just rub a drop between thumb and first finger and close to the sensor to trigger the alarm.

PROJECT 33
ENTRANCE MONITOR (SHADOW CAST)

Function

The entrance way to a place of business or any other location can be monitored by this circuit, which is sensitive to the slightest light change or slightest shadow cast. No direct source of light to the sensor is necessary. The circuit sounded an alarm when a person passed 15 feet in front of the sensor aimed at overcasted daylight that was 30 feet from the unit. (See Fig. 7-35.)

IC1, LM386, is a low-voltage audio amplifier that is connected as an oscillator. The oscillator is disabled by a reference voltage via IC2.

IC2, ULN333OY, is an optoelectronic switch with output pin 1 held to 6.87 volts with incident light on the sensor element. This presents IC1 pin 3 with a reference voltage that disables IC1 from oscillating. However, if a slight shadow is cast on the sensor element, pin 1 goes LOW to about .33 volts and current via R1, D1, and R3 to pin 5 establishes a reference voltage. Also C1 charges through R2 to rise above this reference voltage to make output pin 5 go LOW and the completion of 1 hertz. The frequency rate is about 1 kHz with C1 at .01 value.

The output frequency is coupled to the speaker via C2. Compo-

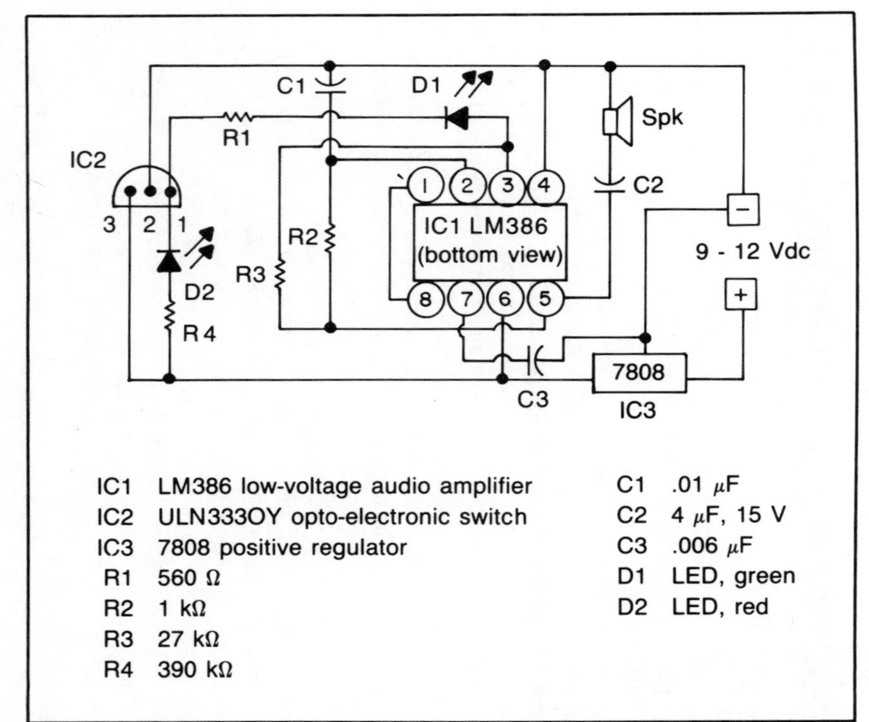

Fig. 7-35. Entrance monitor (shadow cast).

nent C2 is kept small in value to match the energy capability of an 8-ohm, 3-inch speaker and reduced loudness. If a larger speaker is desired for louder tone, increase the value of C2.

The standby current was 10 mA while the operational current was 60 mA. It would be advisable to utilize an ac converter for the power source rather than the battery operation if long period usage is the goal. The IC3, 7808, is necessary to clamp the voltage for stabilization of the frequency and sensitivity. Direct voltage beyond 8.5 volts caused the oscillator to be free running, that was independent of the sensor.

The sensor unit, IC2, should be mounted in a flashlight reflector in reverse position for narrow incident light sensitivity. The focal point adjustment can be found experimentally. Also, room incident and reflected light would be a factor and must be considered when establishing location. Use a mirror for reflected artificial light might be necessary in some instances. LED, D2 is in the circuit to monitor the sensor. The Green LED in the oscillator is part of the reference dividing network in addition to being a monitor of oscillation.

Further Study

This circuit has wide latitude for experimental research. For example, one might use a LM555 as a monostable that would control the ON delay to OFF of the oscillator with sensor as the trigger source for the one shot. The circuit in the present design will sound constantly if one stands in front of the sensor to cast a shadow. Component C1 can be lowered in value to produce a lower tone.

PROJECT 34
JODEL ALARM

Function

The Jodel Alarm (Yodel) is an ear catching sound to get attention. It can be utilized as a startling sound by itself or coupled to an amplifier system as a general alarm. (See Fig. 7-36.)

IC1, LM386, low-voltage audio amplifier is designed as a low frequency multivibrator via C2, R1, R2, R3, and C3 components. The output pin 5 couples the low frequency to pin 5 of IC2 via a voltage dividing network, R4, R5, and C4.

IC, LM555, is the high-frequency astable multivibrator with frequency determined by C5, R6, and R7. The output pin 3 is capable of driving a large speaker with considerable loudness. The frequency of IC2 is affected also by voltage change, therefore the frequency would change with battery wear. This problem can be overcome by regulating the voltage to IC2 via IC3, a three-point regulator. Pin 5 of IC2 receives the low frequency from IC1 and swings the output to give the yodel affect. Changing the value of C5 will change the high to low effect of the sound.

Further Study

Place an oscilloscope between pin 5 of IC1 and ground to measure

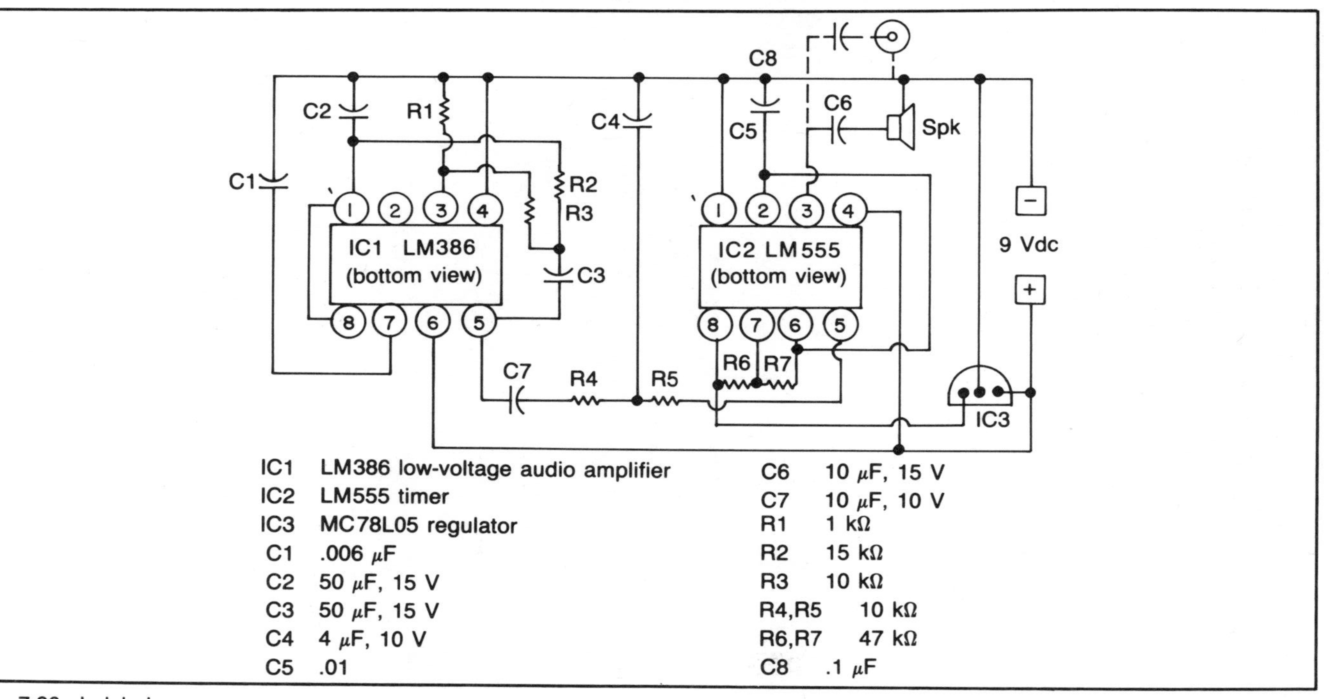

IC1	LM386 low-voltage audio amplifier	C6	10 μF, 15 V
IC2	LM555 timer	C7	10 μF, 10 V
IC3	MC78L05 regulator	R1	1 kΩ
C1	.006 μF	R2	15 kΩ
C2	50 μF, 15 V	R3	10 kΩ
C3	50 μF, 15 V	R4,R5	10 kΩ
C4	4 μF, 10 V	R6,R7	47 kΩ
C5	.01	C8	.1 μF

Fig. 7-36. Jodel alarm.

the swing frequency. Measure the frequency at output pin 3 of IC2 with input pin 5 disconnected to determine the high level. Measure the frequency from IC2, pin 3 with both ICs functioning. Change the value of C5 to .015 μF and note the change in sound. Experiment by placing a second speaker with a 300 μF capacitor between pin 5 of IC1 and ground to hear the combination of sounds. Can you redesign the circuit to utilize another LM555 in place of the LM386 to produce a warble alarm?

PROJECT 35
INFRARED DETECTOR

Function

The infrared detector will sense any infrared light whether incandescent lighting, heating element, flame, or infrared LED. (See Fig. 7-37.)

IC1, TIL 113, is a photo-darlington optocoupler with high-gain characteristics. The slightest detection of infrared by Q1 will turn on Q1 to light the internal LED of the optocoupler, IC1, which is detected by the photo-darlington.

The photo-darlington side, pin 4 is the emitter, pin 5 the collector, and pin 6 the base. The base is not needed because the photo-darlington can be biased on by the internal LED. The emitter resistor R2 from pin 4 is at low or ground when the photo-darlington is in nonconducting state. When infrared light strikes Q1, it conducts to light the internal LED. The photo-darlington transistor is biased ON by the LED and conduction takes place via R2 and R3. The voltage at pin 4 rises to make enabling pin 4 of IC2 to go HIGH.

IC2, LM555, is a multivibrator with a frequency of about 600 hertz as determined by C1, R5, and R6. The oscillator came on when reset pin 4 went HIGH.

A sensitivity control could be connected in place of R1, like a

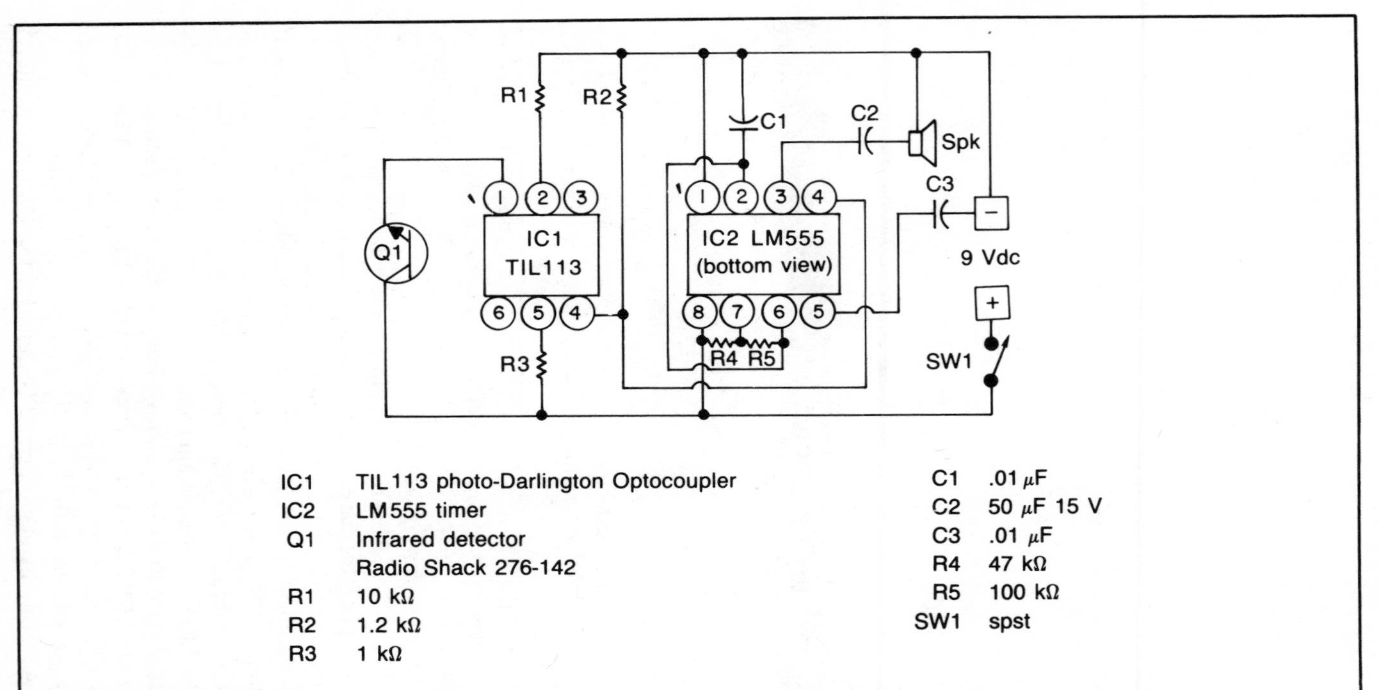

IC1	TIL113 photo-Darlington Optocoupler
IC2	LM555 timer
Q1	Infrared detector Radio Shack 276-142
R1	10 kΩ
R2	1.2 kΩ
R3	1 kΩ
C1	.01 μF
C2	50 μF 15 V
C3	.01 μF
R4	47 kΩ
R5	100 kΩ
SW1	spst

Fig. 7-37. Infrared detector.

1 kΩ limit-resistor in series with a 15 kΩ rheostat. Sensitivity can be controlled also by utilizing the base leg, pin 6 with a limit-resistor and a rheostat to set bias current for a partially ON state by the photo-darlington. The circuit is not latched, therefore when the infrared source is removed the sound will go off.

Further Study

Try different sources of infrared light and different distances from Q1 to determine distance versus light type for trigger capability. This has the possibility for a photoflash slave. Can you think of other uses?

PROJECT 36
INFRARED DETECTOR/ALARM

Function

The security of an object in total darkness can be assured by an infrared invisible light system as the security detector. The object under guard cannot be moved without tripping an alarm. These are often utilized in show case displays. (See Fig. 7-38.)

IC1, CD4009, is a hex inverter IC in which three inverters are utilized as switches. The first inverter, pins 2 and 3, are the detector switch. Pin 3 input is triggered HIGH by the infrared photo detector Q1 via R1, which sets the input to about 1 volt threshold with no infrared source present. An object is placed between Q2 (high output infrared LED) and Q1 the infrared detector. When the object is moved and the beam strikes Q1, the voltage rises to about 7 volts which makes pin 2 go HIGH. Since pin 2 is connected to pin 5, the second inverter switches to make output pin 4 go HIGH and the red LED, D1 turns on.

Pin 7 of the third gate was LOW from grounding via R4, R2, and its output pin 6 was HIGH to turn on the green LED, D2, to indicate everything is secure in the ready state. However, pin 7 went HIGH when D1 turned on via the first inverter and the green LED goes off. The second inverter is in the circuit to latch the output.

IC2, LM555, is a multivibrator of about 625 hertz as determined

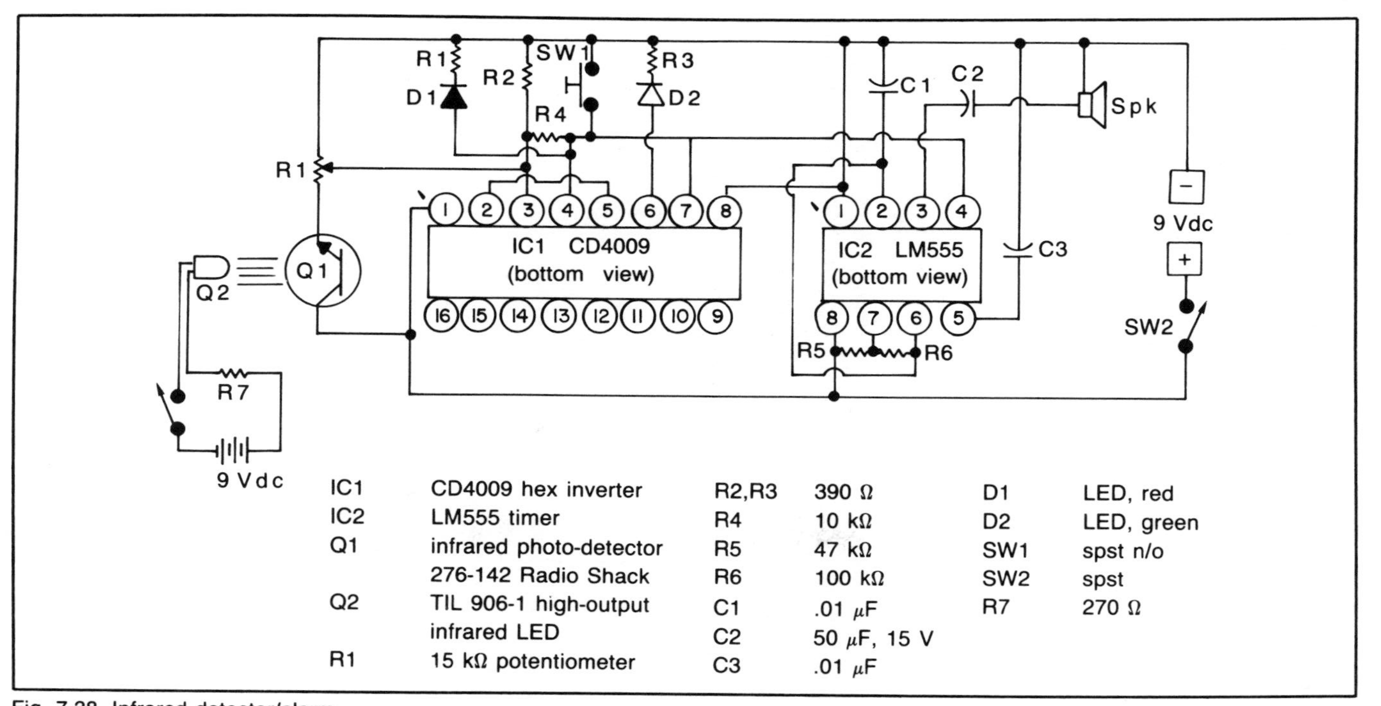

Fig. 7-38. Infrared detector/alarm.

by C1, R5, and R6. The reset pin 4 is held LOW when pin 4 is LOW in the set condition. When pin 4 went HIGH as a result of the first inverter switching, IC2 was enabled and the alarm sound went on.

What one sees in this action is the green LED going off, red LED coming on, and alarm sound comes on. The circuit is latched by the second inverter. If one returns the object to come between Q2 and Q1, and then press reset switch SW1, the circuit is reset because SW1 switches the third inverter LOW to make the output HIGH and the green LED (all secure) comes on. Pin 4 goes LOW and reset pin 4 of IC2 is LOW to keep the alarm off. The red LED is off because pin 4 is LOW via the reset and the object in place.

The standby current is 14 mA of which 4.7 mA is for IC2 in ready state. The 9.3 mA is for the green LED in *OK security*. The operational current is about 160 mA with a 4-inch speaker. The high output infrared LED utilizes 58 mA which provides a distance of 20 inches as effective watch space. The beam is quite narrow in spite of the incoherent light. It takes a little practice to direct it at Q2.

Further Study

Try different size objects on a pedestal to determine how narrow the infrared beam projects. Remember, infrared light is present in incandescent lighting and to some extent in fluorescent lighting. If one points Q1 toward the ceiling lights, the alarm will trigger.

PROJECT 37
TRF AM RECEIVER/AMPLIFIER

Function

The Ferranti ZN414 makes an excellent (TRF) tuned-radio-frequency tuner for coupling to an IC low-voltage audio amplifier to make a complete radio. A minimum of external parts are needed to provide a circuit for simplicity and economy of space. It will function without an earth ground although an earth ground improves sensitivity, particularly in low reception areas. Earth ground will help to reduce interference and other noise factors. (See Figs. 7-39 and 7-40.)

IC1, ZN414, is a multistaged transistor array and detector circuit. The voltage at output pin 1 via R2 is critical and should be held constant otherwise distortion will occur. The circuit was designed to clamp the voltage at about 1.23 volts at the junction of R3 and R5, which in turn kept the voltage constant at pin 1 via R2. A three-point positive voltage regulator, 78L05, held the dc level constant to the voltage divider network R3 and R5 even when the supply voltage shifted from 6 to 12 volts under test. The current draw from the ZN414 was negligible at .26 mA. The output is fed back to the input via R1 to provide agc action. The circuit will not function without C5, which is the rf return path for the detector stage. (See Fig. 7-40.)

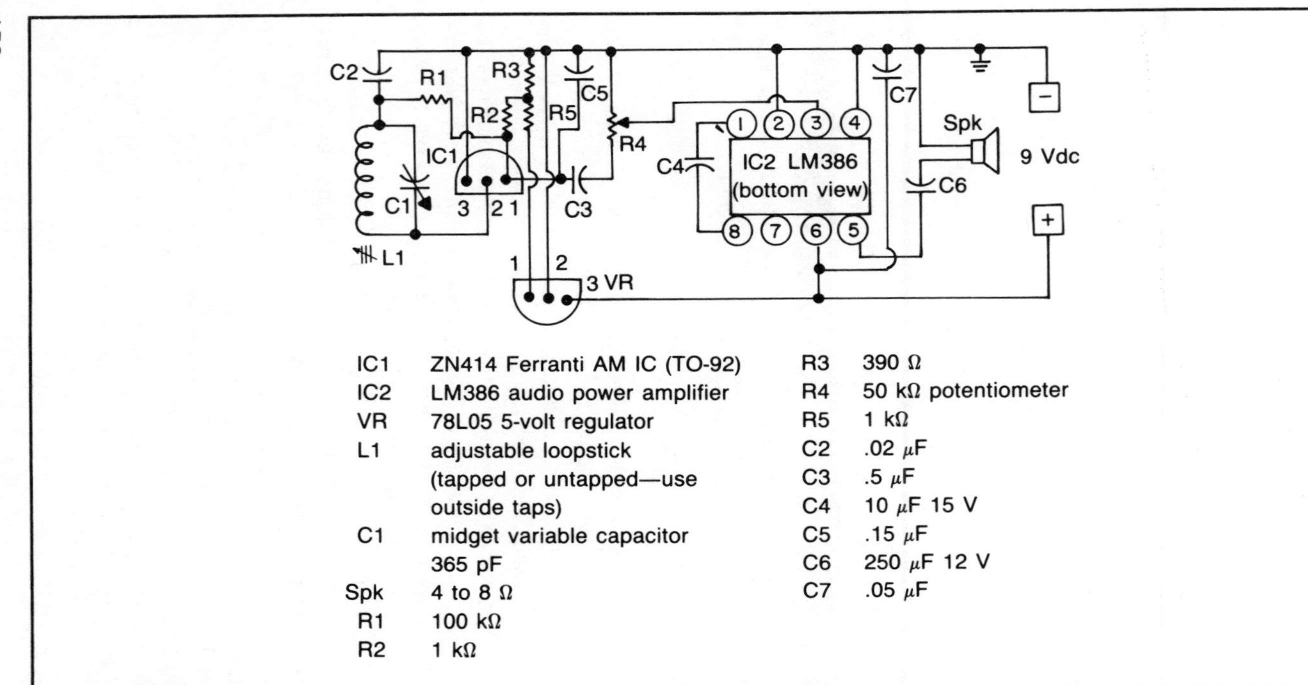

IC1	ZN414 Ferranti AM IC (TO-92)	R3	390 Ω
IC2	LM386 audio power amplifier	R4	50 kΩ potentiometer
VR	78L05 5-volt regulator	R5	1 kΩ
L1	adjustable loopstick (tapped or untapped—use outside taps)	C2	.02 μF
		C3	.5 μF
		C4	10 μF 15 V
C1	midget variable capacitor 365 pF	C5	.15 μF
		C6	250 μF 12 V
Spk	4 to 8 Ω	C7	.05 μF
R1	100 kΩ		
R2	1 kΩ		

Fig. 7-39. TRF AM receiver/amplifier.

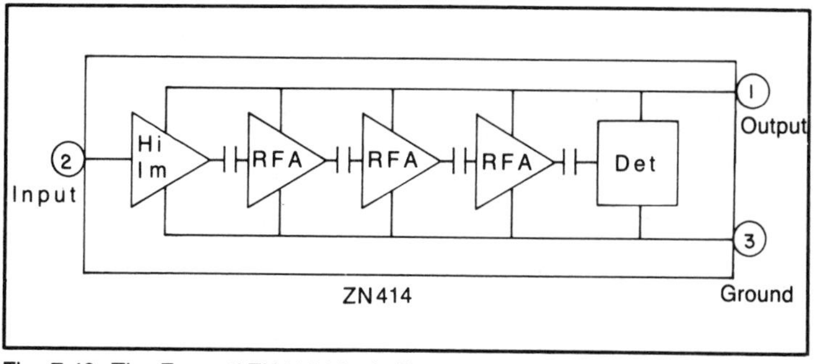

Fig. 7-40. The Ferranti ZN414 block diagram.

The output from IC1 is coupled through C3 to ground via R4. This voltage division of the output of R4 provides volume control via input drive to IC2. IC2, LM386, is a low-voltage audio amplifier with ample drive to the speaker. Various size speakers will work but larger sizes are more efficient and will produce better quality. Standby current was from 15 to 18 mA at low volume levels and went over 100 mA at larger volume settings. The LM386 has a tendency to self-oscillate in proto-board layout and sometimes in an etched circuit board. Decoupling of the dc with capacitor C7 from pin 6 to ground is essential to avoid this problem.

Further Study

Remember a TRF tuner will tune broadly: it will be possible to hear more then one station at a time. Try winding your own coil; about 41 turns of litz wire on a 1/4-inch form to a length of 1 inch should work, or wind it on a ferrite rod.

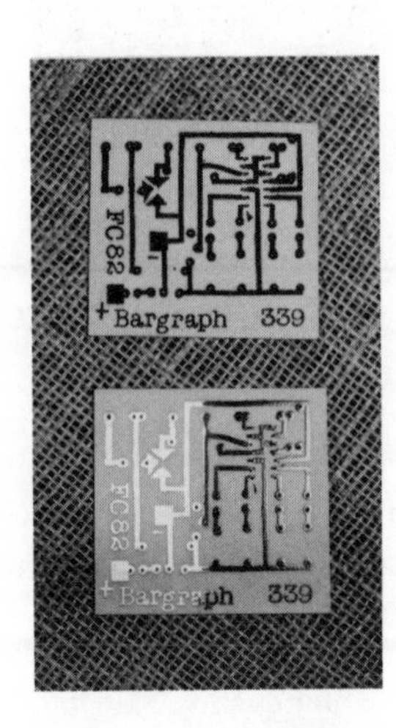

Chapter 8

Light Control Projects

PROJECT 38
FAIL-SAFE LIGHT

Function

A fail safe light is useful as a standby device that turns on in the event of ac power failure. It is most useful in places that would be in total darkness in the event of a power failure.

The power company supply of energy is monitored by T1,D1,R1,C1,C2, and R2, which is a converter for changing ac to dc. The converter produces approximately 8.9 volts (6.3×1.414) that is coupled to the enabling pin 1 and 2 of IC1. (See Fig. 8-1.)

IC1, CD4071, is a quad 2-input OR gate with a truth table that shows output pin 3 HIGH when either inputs are HIGH or both inputs HIGH. The output is connected through R4 to Q1 base of a pnp transistor. This positive voltage places Q1 in reverse bias and the transistor is in cutoff, or the light is off. The light will come ON if the base of Q1 is negative. When the power company energy is present, IC1 output pin 3 is on (or positive) and Q1 is off. If the power goes off, pin 3 is low and Q1 conducts and thus turning on the light.

When Q1 is on, the current drain is approximately 130 mA, which would be severe for a 9-volt transistor-radio type battery. But this light is for emergency use and will be used for a short time until other arrangements can be made. Standby energy on the converter is about

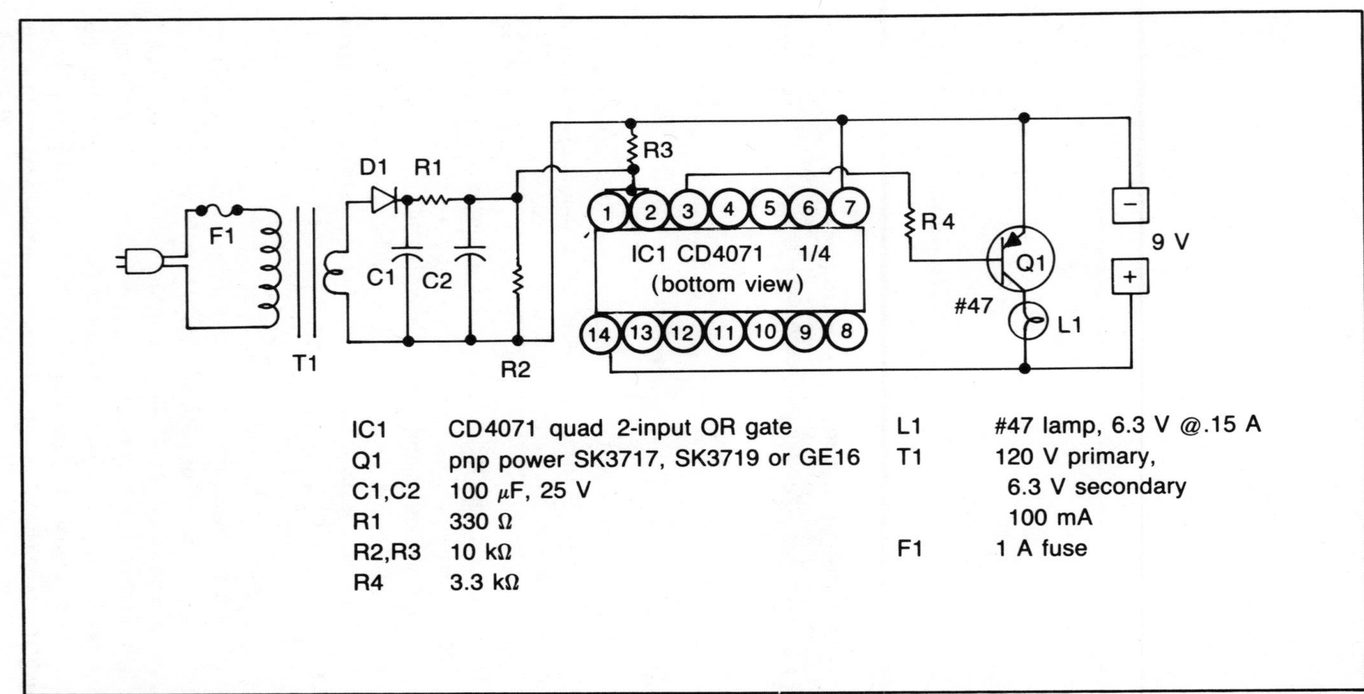

Fig. 8-1. Fail safe light.

2 mA so operation cost for standby is nil. Checkout is simulated by pulling the plug on the transformer to show power loss.

Further Study

Try an AND gate, CD4081, in place of IC1. Its truth table shows output HIGH when both inputs are HIGH. Can you think of a way to keep a ni-cad battery on charge from the ac converter and be the supply for the circuit in the event of power failure?

PROJECT 39
DARKROOM IN USE WARNING LIGHT

Function

Light sensitive paper and film can be ruined by someone inadvertently entering a darkroom when processing is in progress. The circuit here reads the situation inside the darkroom and warns others whether they may enter or not.

IC1, LM555, is a low-pulse multivibrator with a pulse rate of 2 per second with a positive level of .2 second duration. Changing the value of C1 can change this frequency. The output pin 3 is coupled through R3 to the gate of a triac. The duration of the positive pulse to the gate of the triac is short enough that the triac is not fully gated on. The ON time produces about 60 volts across the 40-watt lamp so the lamp is half of the rated power and brightness. On a very positive pulse the triac turns ON and the ac frequency at the triac commutates or turns off during the negative duration of the multivibrator. (See Fig. 8-2.)

IC1 oscillates when pin 4 is positive and stops when it is negative or low. The components R4 and LDR (light dependent resistor) is a voltage-divided network across the dc supply. The ratio of the resistance (and voltage drop) between R4 and the LDR determines the positive or negative state of pin 4. When light is on the LDR, its re-

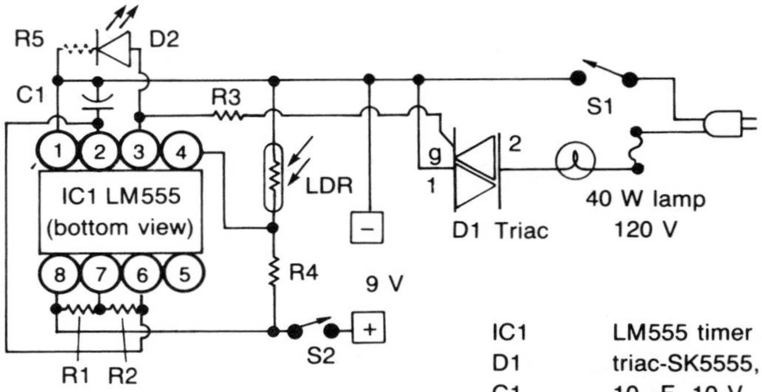

IC1	LM555 timer
D1	triac-SK5555, IGT 5 mA, 200 DROM
C1	10 μF, 10 V
R1,R2	100 kΩ
R3	1.8 kΩ (at this value light flashes at half brightness
R4	330 kΩ
LDR	light dependent resistor 2 kΩ to 1 MΩ
S1,S2	spst
F1	1-A fuse
R5	470 Ω
D2	LED

Fig. 8-2. Darkroom in use warning light.

sistance is low or about 2 kΩ and since R4 is fixed at 220 kΩ, the voltage at pin 4 is low and the oscillator is OFF. When the LDR is in darkness the resistance is about 1 megohm, the voltage at pin 4 is high and the oscillator is ON and the lamp flashes.

The unit is placed on the outside of the darkroom entrance, but the LDR is extended inside to read the room light condition. If a white light is to mean 'OK to Enter,' the outside red light is off. If the room is dark, the light is flashing 'Warning-Do not Enter.' Standby current is 5 mA; operating current (light flashing) 10 mA.

Further Study

Increase the value of C1 to change the pulse rate. Change values of R1 or R2 to change the duration time of OFF or ON. Why add the LED? What purpose would the LED serve?

PROJECT 40
NUMBERS SPINNER/ROULETTE

Function

Number spinners are useful in games and in other chance choices. This circuit is a true electronic wheel in that the numbers rise slowly to full swing and then descend at a slower rate to stop on one position.

IC1, SCL4046, employs a voltage-controlled oscillator at a pulse rate of 40 hertz at threshold by R6 and C2. The pulse rate is controlled by varying voltage at pin 9. When S1 is closed, C1 charges slowly through R2 thus producing a rising voltage at pin 9. The rising voltage causes the frequency to rise from zero to threshold and remain at that frequency as long as S1 is closed. When S1 is opened, C1 discharges slowly through R1 to ground and the voltage falls toward zero, thus producing a decreasing pulse rate. The output pin 4 is coupled to the clock input of IC2 pin 14 via C3. (See Fig. 8-3.)

IC2 decoder advances through each output (0-0 to 0-9), pins 1 through 7, and 9 through 11 with each input pulse of IC1. IC2 has ten outputs in sequence as shown in the pinout diagram. The outputs go HIGH at each 'turn on,' which causes each LED to light. Only eight outputs are utilized in this circuit thus giving two numbers to the spinner or the house. One can arrange the lighting order in sequence or

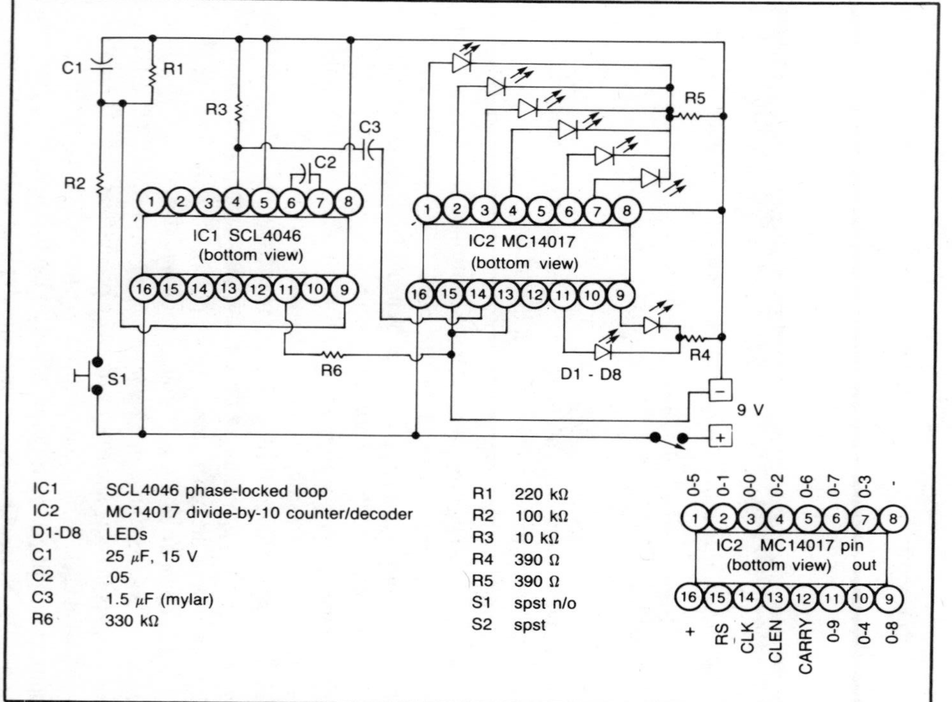

Fig. 8-3. Numbers spinner/roulette.

staggered combinations. They may be grouped in a circle, straight line, or some other configuration.

Further Study

Can you change the diodes to reverse the process (all ON and the OFF stop is the choice)? What is the standby current? Can you change components to vary the up-side and down-side rates?

PROJECT 41
SAFETY FLASHING LIGHT

Function

An amber flashing light can be used as an emergency warning. It could be utilized in addition to the 4-light flasher on an automobile when changing a tire along a busy highway, on a curve of the road or over a crest. It can be plugged into a cigarette lighter and hung out of the window on the traffic side or set on the roof.

IC1 is a timer operating at a rate of two pulses per second that is set by C1, R1, and R2. The time ON can be calculated; $t_1 = .7 (R1+R2) C1$ or $t_1 = .7 (100 k+100 k) .00002 = .28$ seconds. The output is not symmetrical therefore the OFF time is of shorter duration than the ON time; $t_2 = .7 (R2) C1$; or $t_2 = (.7 (100 k) \times .000002$, or about .14 seconds. (See Fig. 8-4.)

The pulse is coupled to an npn transistor, Q1, which is a power-switching type. The bias resistor, R3, is set to turn on L1 with an over-voltage of about 8.5 V and over-current of about .18 A for a short duration. This provides a higher intensity flash but will reduce the life of the lamp over time. The lamp could be housed in an amber pill box for a wider area flash surface.

Q1 has a collector current rating of 7 amperes maximum: the .18 ampere switching current will have little effect on the transistor

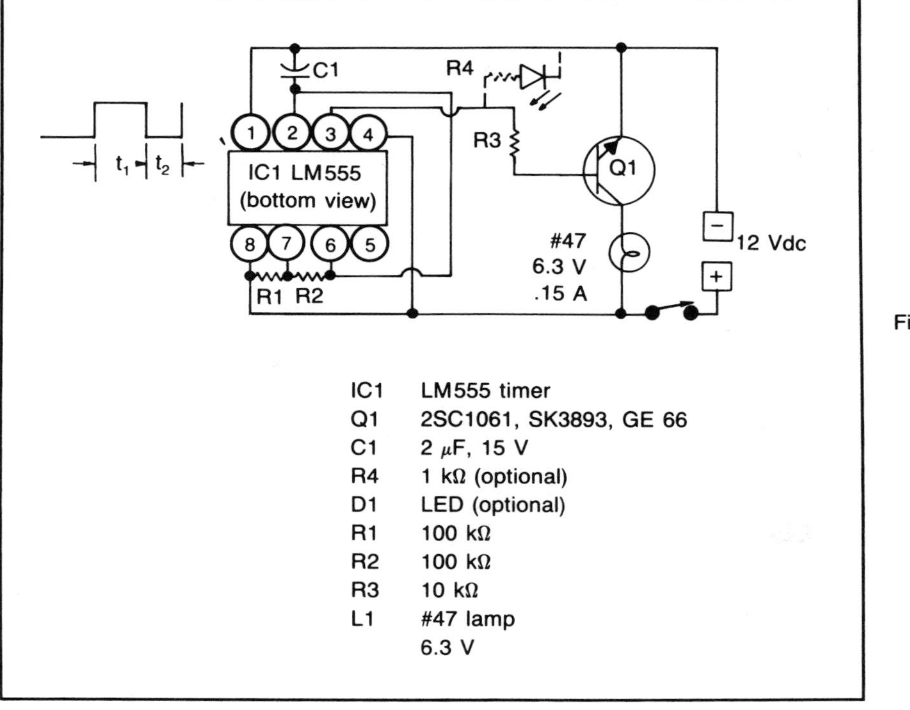

Fig. 8-4. Safety flashing light.

(heatsink if you wish). The bias current through R3 is about .9 mA which is nil for the LM555.

Further Study

Change the speed rate by changing C1: like 1 μF. Change symmetry by changing value of R2; like 220 kΩ. Can you change the lamp and R3 for compatible output; like a #44 lamp and R3?? You might prefer an automotive type lamp for greater intensity.

PROJECT 42
OBSTRUCTION WARNING (NIGHT FLASHING)

Function

This circuit automatically turns ON a flashing light at night and off during the daylight. It monitors a dangerous situation such as a hole, fresh concrete, or other obstruction.

IC1 is an astable oscillator at a frequency of about 6 hertz per second that is determined by C1. The output is almost symmetrical, therefore the ON and OFF times are nearly equal in duration. Standby current (no flashing light) is about 7 mA to which one must add .85 mA bias current when light is flashing. The 7.85 mA operational current is far below the 200 mA maximum for a LM555. The output is coupled through R3 to turn on Q1 with every positive pulse transistion of IC1. (See Fig. 8-5.)

The oscillation of IC1 is controlled through reset pin 4 which turns ON the oscillation when high and OFF when low. The LDR and R4 provide a variable voltage dividing network that is connected to pin 4. When light strikes the LDR, it goes down in resistance making voltage across it to go low, and thus shutting down IC1. When dusk and nighttime comes, the LDR rises to a greater resistance value than R4, therefore pin 4 is positive and the flasher comes on.

Q1 is a power switching transistor with 7 ampere collector

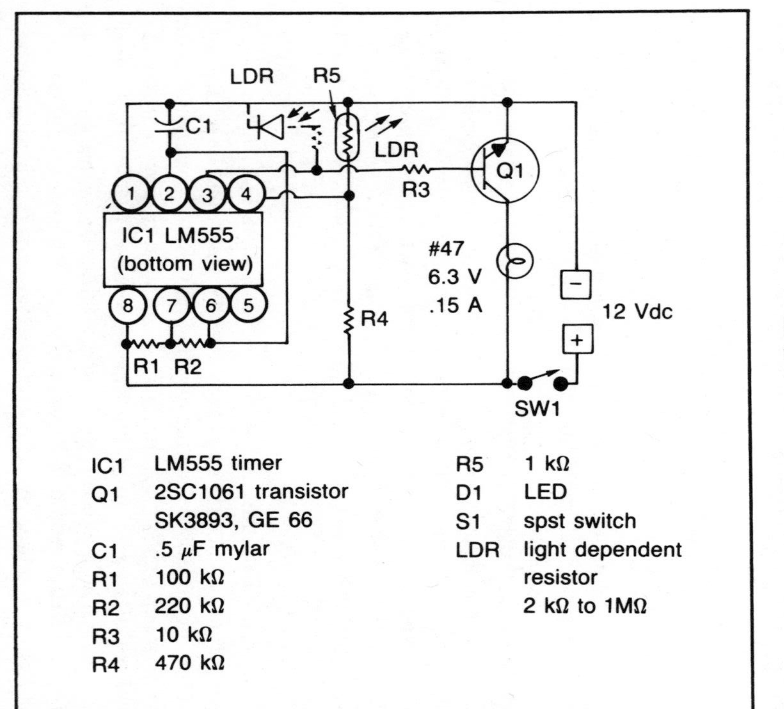

Fig. 8-5. Obstruction warning—night flashing (automatic turn ON and OFF).

capability. It has no difficulty in handling the over-current of .18 amperes of the lamp. The over-voltage and over-current effect on the #47 lamp (.15 rated current) will appear as dark residue at the end of the lamp after extended use. This is a small penalty to pay for the extra brightness of deliberate over-voltage.

Further Study

Increase the value of R3 to reduce the lamp to within tolerance range of 6.3 V, .15 A. Try an amber colored pill container as a reflector.

PROJECT 43
BOAT MOORING LIGHT (AUTOMATIC)

Function

Boats at anchor in navigable waters are required to have a mooring light. This circuit will turn on a light automatically at night and off during the daylight.

IC1, ULN33OY is an optoelectronic switch that features on-chip a photodiode, amplifier, trigger, power driver, and regulator. (See Fig. 8-6.) Maximum energy is 50 mA and 15 V, which is more than adequate for our purpose here. When light strikes the face side of IC, it is in the off state and thus opening the path to ground, pin 2.

Transistor Q1 switches on with a base current of .8 mA to produce an output current at the collector of about 160 mA that makes L1 come to full brightness. The base current is controlled by R1 via the optoelectronic switch to ground. When the optoswitch is off, the voltage at pin 1 is the same as the supply voltage therefore reverse bias voltage for Q1 and it is in cut-off state. When light is blocked from the face of the optoswitch, it turns on providing a path to ground for the base bias network. Q1 comes on when light is screened from IC1.

LED and R2 are in the circuit to monitor the switching; they may be omitted. Standby current in daylight is about 9 mA, the LED on

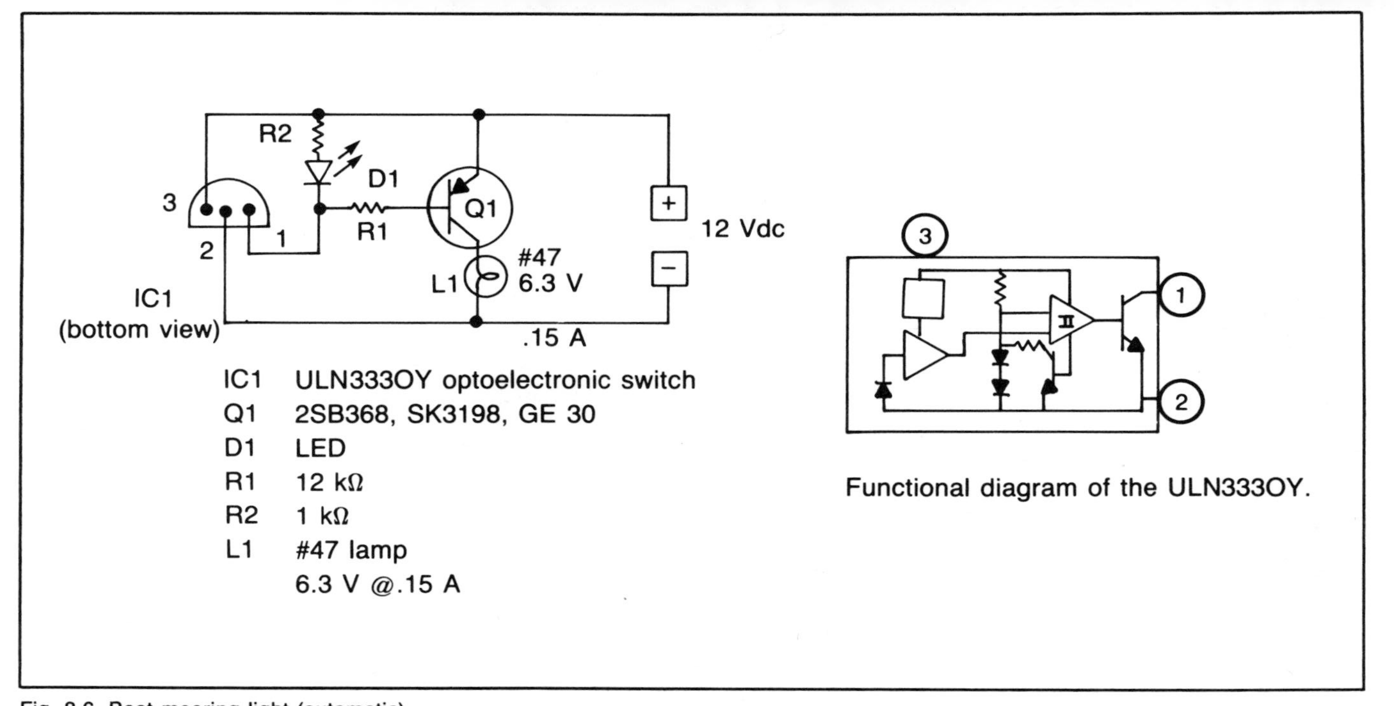

Fig. 8-6. Boat mooring light (automatic).

state is 10 mA, while lamp on current is 150 mA. The total current in night operation is about 169 mA. One could expect in 10 hours of darkness to utilize 1.7 amperehours of energy. This would provide better than 2 months service with a fully charged marine battery.

Remember, IC1 must be shielded from L1 light source or feedback will shut it down. You may have to place IC1 in a shroud and direct it away from L1 or any other artificial light source.

Further Study

Can you use an npn transistor for Q1 by interfacing it with a pnp signal type to produce the same results? Lamp L1 could be housed in a white pill container for weather protection.

PROJECT 44
OCCULTING LIGHT (LIGHTHOUSE SIMULATOR)

Function

An occulting light is a type that is partially on, switches to full brightness, and then decays slowly back to partially on or threshold level. It simulates the action of a lighthouse. It could be utilized for an obstruction light or an automotive flasher.

IC1 is a slow-speed astable multivibrator at a pulse rate of 1 hertz per second as determined by C1. The output pin 3 is coupled to the base of Q1 via R4 and R6. Components R3 and D1 monitor the pulse by turning on the LED when pin 3 is low. (See Fig. 8-7.)

Transistor Q1 is set to turn on a light at a threshold voltage of about 5 volts. This is accomplished by biasing Q1 ON through R5 and R6; increasing the value of R6 will decrease the threshold light level.

When IC1 output at pin 3 goes low during the negative excursion of the pulse, additional bias current is added to Q1 causing it to have greater conduction and thus brighter lamp output. Also, C2 charges to a positive voltage of about 2 volts. When IC1 output at pin 3 goes HIGH, the bias to Q1 is reversed thus cutting back toward threshold; however, C2 discharges slowly through R6 to base-to-emitter providing a decaying bias current and a slow decay in the light toward threshold.

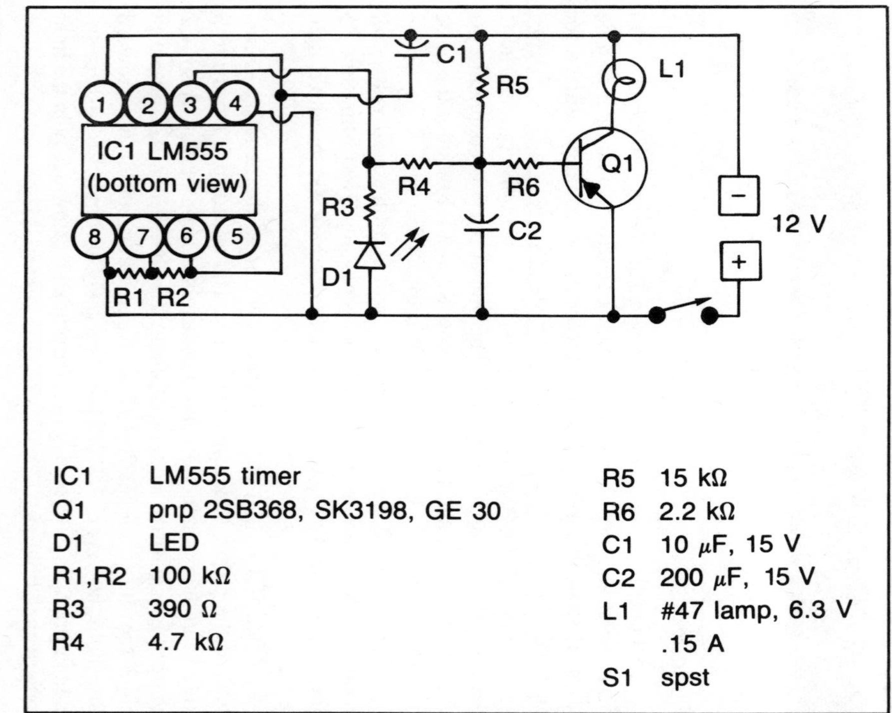

Fig. 8-7. Occulting light (lighthouse simulator).

Further Study

Change the value of C2 to 350 μF and note the difference in decay time. Lower the value of R2 to 47 kΩ and note the time on change. Change the value of C1 to 20 μF and note the astable multivibrator rate. Can you scope the waveform at pin 3 and determine the time ON and OFF periods? Remember, any changes that keep the #47 lamp on for a longer duration (at brightness) will reduce its life since it is working at an overvoltage.

PROJECT 45
DUAL ELEMENT SEALED BEAM FLASHER

Function

A dual-element sealed-beam taillight makes a unique flasher as a warning device. The Tung-Sol 4425R lamp is a 12-volt light with a common input to each element, which draws about 1 ampere each.

IC1, LM555, is a multivibrator of 1 hertz per second. This IC serves as a clock for IC2, dual JK flip/flop. With every positive transistion of the clock input to IC2, flip-flop side pin 1 through 7, the outputs Q and $\overline{Q}$ (not) change polarity. When output pin 2 goes positive, it drives the base of Q1 positive via R5. This transistor switches one element ON. It takes three positive transition pulses from the clock to repeat the ON cycle for element 1 of the lamp. (See Fig. 8-8.)

Pin 1 of IC2 couples its positive pulse to the other flip/flop, pin 13 via C3. When pin 15 goes positive, it drives the base of Q2 positive via R6. This transistor switches element #2 ON. It takes three positive transition pulses from pin 1 of the first flip/flop to repeat the cycle. Note that R6 is larger than R5, therefore element 2 will not be as bright as element 1. What one sees is a differential in light intensity.

The total lighting cycle is (1) all elements out, (2) low element comes ON, (3) delay time for low element ON, (4) high element comes

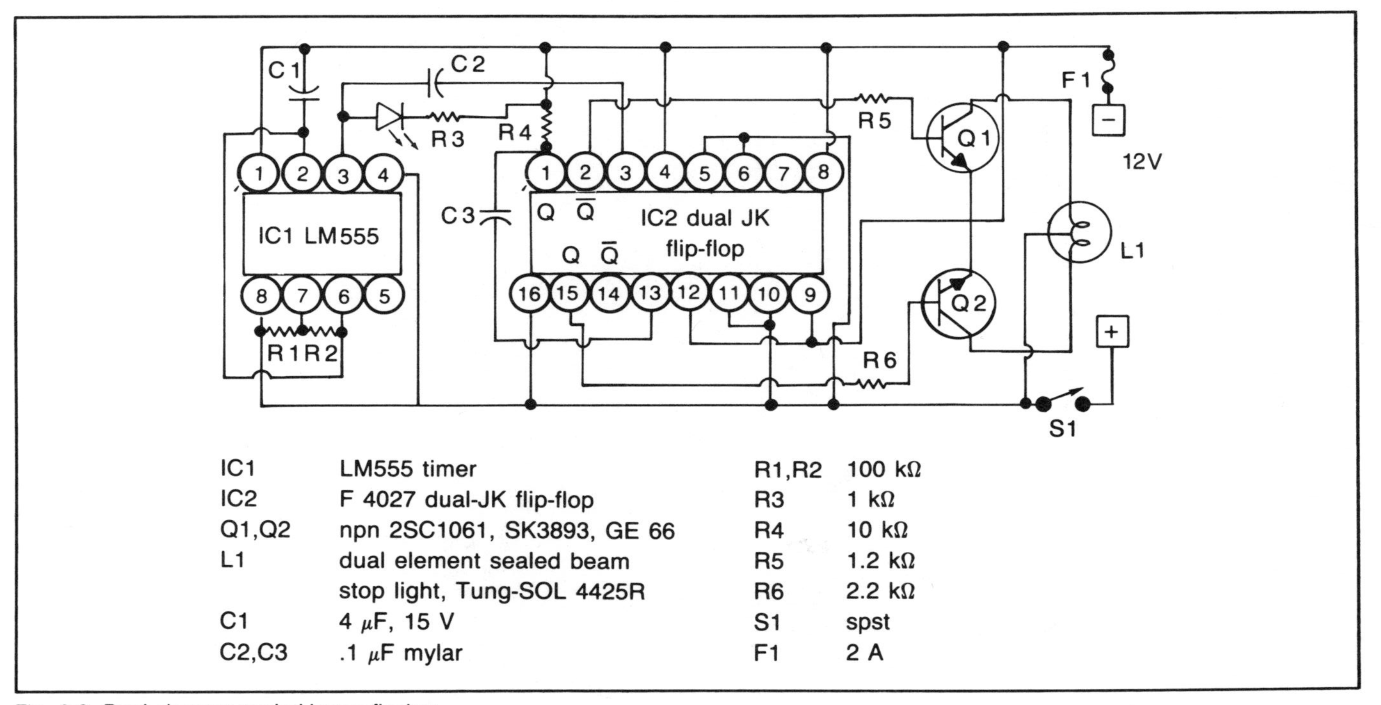

IC1	LM555 timer	R1,R2	100 kΩ
IC2	F 4027 dual-JK flip-flop	R3	1 kΩ
Q1,Q2	npn 2SC1061, SK3893, GE 66	R4	10 kΩ
L1	dual element sealed beam stop light, Tung-SOL 4425R	R5	1.2 kΩ
		R6	2.2 kΩ
C1	4 μF, 15 V	S1	spst
C2,C3	.1 μF mylar	F1	2 A

Fig. 8-8. Dual-element sealed-beam flasher.

ON as an override, and (5) all elements OFF. One sees an occulting affect except for complete shut down.

The low and high elements draw .6 and .9 ampere respectively or 1.5 amperes when both are on simultaneously. Bias to the transistors is between 3 and 5 mA. The transistors should be heatsinked in continuous duty to prevent thermal gain effect.

Further Study

Change the value of C1 from 4 μF to 2 μF. Change the output from the second flip/flop pin 12 to $\overline{Q}$ (not) at pin 14 and note any lighting cycle change.

PROJECT 46
SIMPLE FLASHER/BATTERY MONITOR

Function

A lamp flasher utilizing a single IC is inexpensive to construct and will change flash rate with a change in supply voltage. It might be utilized as a warning light that also tells the condition of a battery by observing the flash rate. (See Fig. 8-9.)

IC1 is a transistor array that utilizes an external capacitor for feedback to start and sustain oscillation. The output pin 2 drives a #47 lamp directly. In addition, it drives D1 via R1 and pin 6 that is connected internally to the supply voltage via an internal resistor.

The pulse rate is six pulses per second with a 9-volt supply and components shown; and two pulses per second with a 6-volt supply. Changing the value of C1 will also change the frequency (increasing the value will lower the frequency).

Further Study

Change the value of C1 to 400 μF and note the change in frequency. Vary the input voltage from 1.5 volts to 9 volts and note the change in frequency.

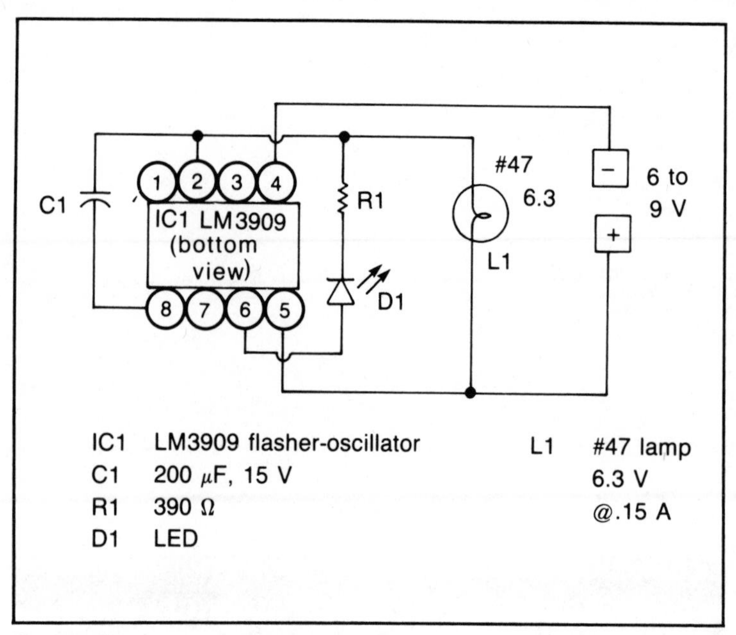

Fig. 8-9. Simple lamp flasher/battery monitor.

PROJECT 47
PAINTER MARKER (FLASHING)

Function

A painter is a float that holds the end of a line from the buoy when a boat is away from the mooring. A low drain flashing light to identify a painter at night is useful. The circuit features an additional interface for automatic shutdown in daylight. It's great for daylight departure and nighttime return.

IC1 is a flasher-oscillator that drives two LEDs in parallel from the output at pin 2. Two additional LEDs are driven in the feedback side via pin 6. The LEDs are situated at 90 degree points to be visible from several approaches. The flash rate is determined by C1; increasing its value will decrease the flash rate. The flash rate is about 6 pulses per second with a 470 μF capacitor as shown. The circuit shown here (solid lines) is free running. (See Fig. 8-10.)

Removing the jumper wire between pin 1 and 8 causes the shut-down of oscillation. A transistor with emitter-collector leads replacing the jumper wire will shut-down when cutoff and turn-on when saturated. A 10 kΩ bias resistor connected between base and positive will turn the oscillator ON as though the jumper wire was reinserted. Adding a light-dependent resistor between base and negative (dashed lines) will reduce the bias current from Q1 in daylight. The bias cur-

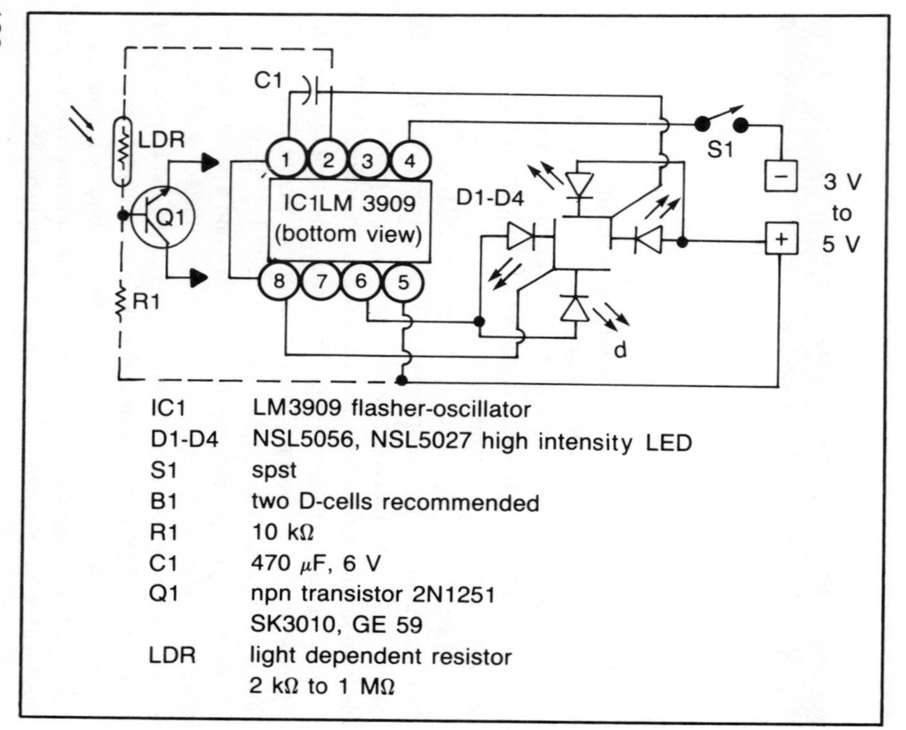

Fig. 8-10. Painter marker—flashing.

rent in the flashing state is .11 mA and .01 mA at shutdown during daylight. The flashing current load is low (about 10 mA) while standby current is about 3 mA.

Further Study

Stack four more LEDs in parallel to increase visibility. Change the value of C1 to change the flashing rate.

PROJECT 48
WINDOW COMPARATOR/TEMPERATURE MONITOR

Function

A monitor of temperature at two levels is possible with a window comparator. This type of circuit would be valuable in photography where operating temperatures of a chemical must be maintained between two temperatures.

IC1 is a dual operational amplifier with each side acting as a comparator; however, a single thermistor is the common sensor. Side A is the first ON side with R2 and R3 as the reference voltage at pin 5. The voltage at pin 5 is set below the voltage at pin 6, which comes from the voltage dividing network R6,R7, and R9 of side B. When thermistor R9 becomes warmer, the resistance falls and the voltage across it also falls. When it falls below the reference voltage on side A, the LED, D1 (yellow) will light. The reference voltage on side A is set to 1.6 volts and the resting voltage at pin 5 is 2 V. (See Fig. 8-11.)

Side B, pin 3 has a reference voltage at 3.8 or below the resting voltage at pin 2, which was set at 4.2 volts. When the thermistor becomes warmer, the resistance falls and the voltage at pin 2 falls. When it falls below the reference voltage at side B, the red LED will light. Remember side A is sensing the same change and LED (yellow) has already been ON.

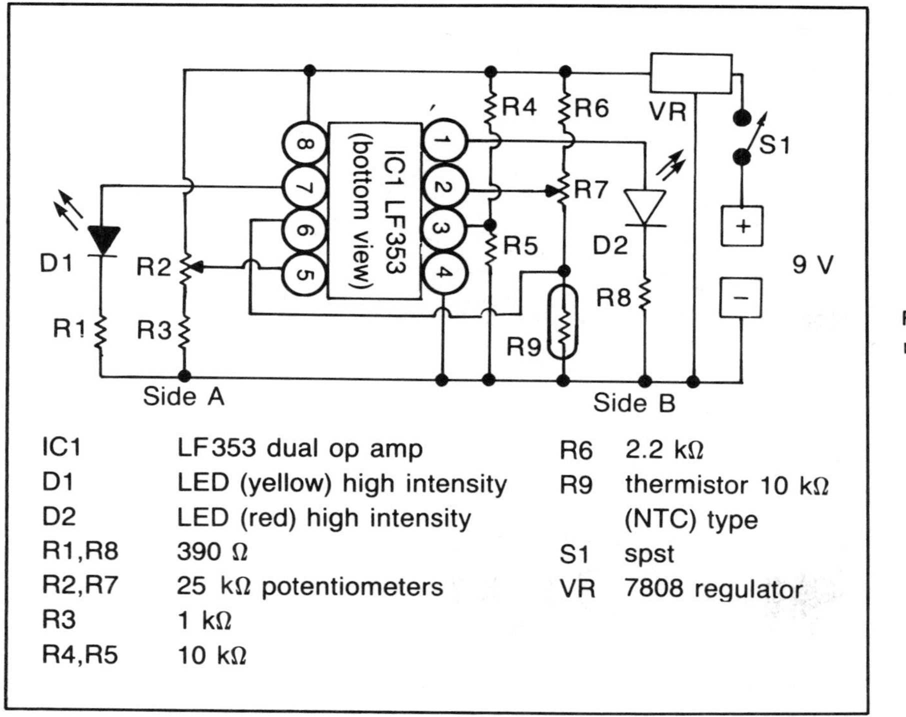

Fig. 8-11. Window comparator—temperature monitor.

The differential between D1 and D2 turning ON is determined by adjustments of R2 and R7. Note in side A the reference is adjustable at pin 5 via R2, whereas in side B the reference is fixed via R5 and R4. The threshold of side A is stable when thermistor R9 has constant temperature, whereas pin 2 of side B is adjustable when R9 is at constant temperature. Adjusting R2 and R7 determine the differential of the window.

It is advisable to read the voltage at pin 6 first, and then at pin 5 when adjusting R2 to determine turn-on and the voltage differential that will be seen at the thermistor in order to make a change. Do the same measurements at pin 2 and 3 of side B when adjusting R7. Testing can be accomplished by heating a paper towel against a register and then pressing it against the thermistor.

Further Study

Make adjustments with R2 and R7 while reading voltages at the reference and the input voltages. Can you connect a LM555 oscillator into the circuit to sound an alarm when the upper window LED comes ON?

PROJECT 49
STEREO BALANCE VU DISPLAY

Function

The stereo balance VU display is a circuit to balance the stereo outputs, which is necessary for good stereo effect. A mono signal is injected into the amplifier and the outputs to the speakers are applied also to the two inputs of the VU display. Balance control is adjusted to provide an equal number of LEDs that are on in each side when balance has been achieved. (See Fig. 8-12.)

IC1 an IC2 are two quad comparators in a balanced design. Inputs pins 4, 6, 8, and 10 are achieved via Q1 and Q2 as cathode followers. This arrangement was planned to permit considerable drive before lights would come on and to decrease the effect of loading on the output terminals of the amplifier. The 10 kΩ resistors in the voltage reference points of the comparator, pins 10, 8, 7, 5, and positive supply, provide voltages of 1.8, 3.6, 5.4, and 7.2. The increments should be equal for a good linear sequence of lights on. It is important for all 10 kΩ resistors to be of equal value for accuracy of turn-on. You could use precision resistors but this is unnecessary if you measure many resistors and select the matched types.

Transistors Q1 and Q2 should be matched for similar characteristics. The circuit requires 10 V P-P to turn on the first pair

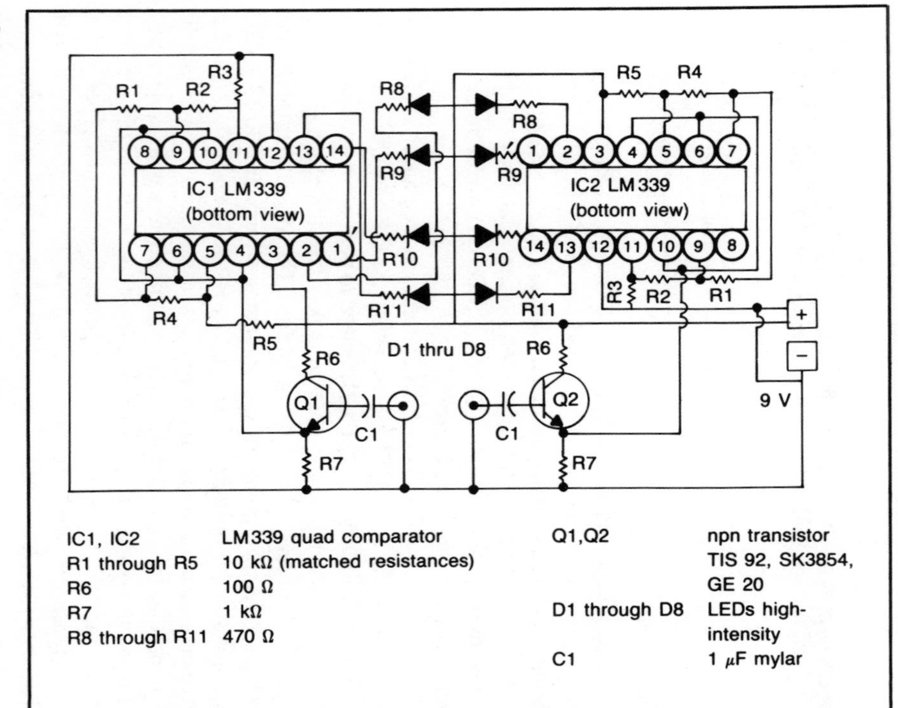

Fig. 8-12. Stereo balance VU display.

of LEDs, 15 V P-P for second pair of LEDs, etc. The LEDs will come on from the bottom to the top direction. If an audio generator of constant tone is used for the signal injection, the LEDs will come on in sequence as the volume control is advanced. If an on-the-air program or record is utilized, the LEDs will flicker back and forth proportionately to the signal level. You can balance with either source, however the amplifier must be set for monophonic. If a monophonic set switch is not available on the amplifier, you must inject the signal to both amplifier-channel inputs at the same time.

Further Study

Raise the value of R5 from 10 kΩ to 27 kΩ and note the change in LED turn-on spacing.

PROJECT 50
STROBED COMPARATOR (TEMPERATURE MONITOR)

Function

A strobed comparator is valuable for monitoring a condition such as temperature-limit by stopped action of a strobed output when the limit is reached. Comparator LM311 is unique in that its output can be strobed in addition to a comparative measurement. (See Fig. 8-13.)

IC1, LM311, pin 3, is the referenced voltage point for the comparator via R1 and R2, which are equal in value. Thus, the reference voltage is 4 volts with a regulated 8-volt supply. The variable voltages at pin 2 are set via the R3 potentiometer and the thermistor. When this voltage is set above the reference voltage, the output pin 7 is LOW and the LED is on. When heat strikes the thermistor, its resistance decreases and the voltage falls below the reference voltage and the output pin 7 goes high, thus the LED goes out. What one sees is a go and no-go state or limit; in this circuit we use a green LED. When the green LED was on the temperature was 64 degrees F and when the LED went out the temperature was 75 degrees F, thus one sees an upper limit monitor.

IC2 is a slow pulse multivibrator of 1 pulse per second. Output pin 3 is connected to the base of an npn transistor, which is the strobe interface of LM311 comparator via pin 6. The green LED from the

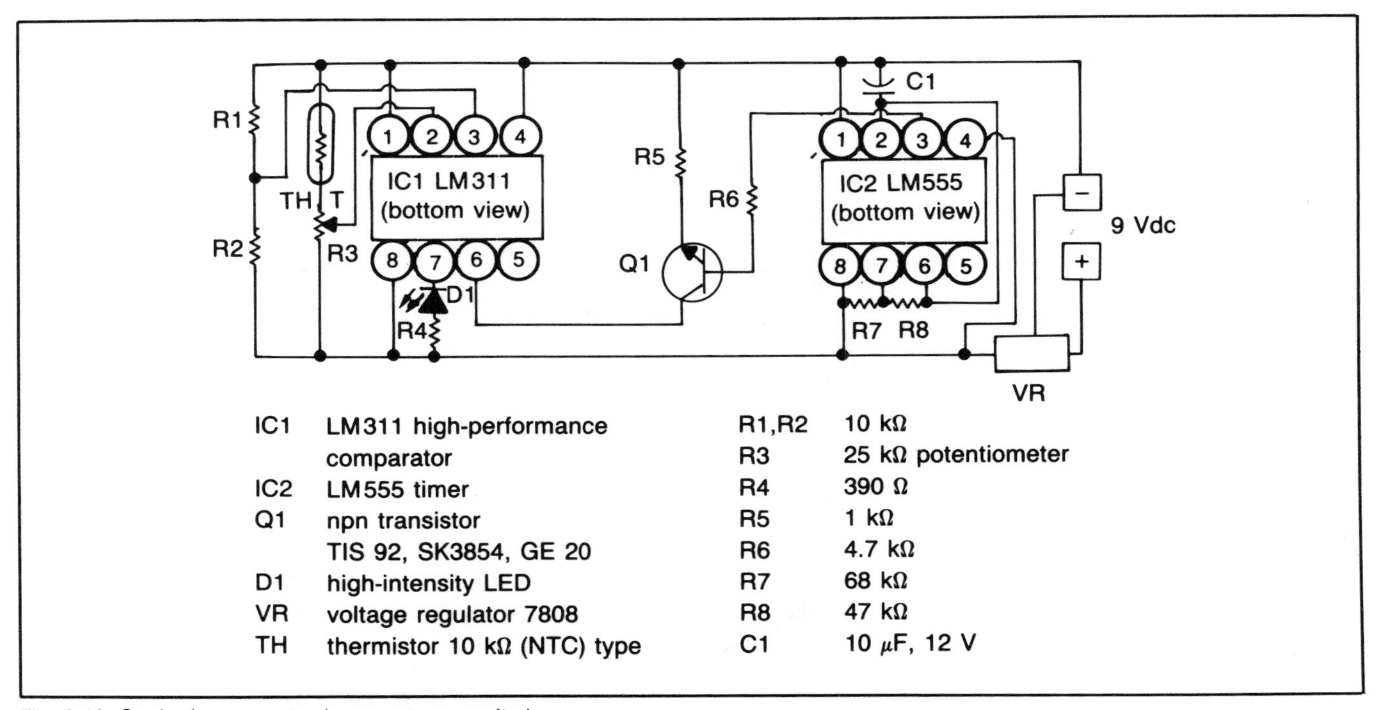

Fig. 8-13. Strobed comparator (temperature monitor).

output of LM311 flashes at the rate of one pulse per second when the comparator is being strobed. What one observes is a flashing green LED that signifies all is OK or within limits. When the flashing light is out, it signifies or communicates that the upper limit has been reached. The differentiating range is adjustable via R3.

Further Study

Can you connect a red LED in conjunction with a pnp transistor and output pin 7 to come on when the green-flashing LED goes out?

PROJECT 51
BARGRAPH MEASUREMENT
(TEMPERATURE, LIGHT, HUMIDITY)

Function

A bargraph can measure temperature, light, humidity, sound, and pressure through an LED display. One can go further to have the circuit activate another IC to perform another task like, set off an alarm, turn on a fan, shut-down a heater, and similar possibilities. (See Fig. 8-14.)

IC1, quad comparator, is connected to make each comparator in cascade arrangement. The reference voltage level for each comparator is set at pins 11, 9, 7, and 5 via R3. The dividing resistors between ground and these pins produce voltages of equal spacing like .5, 1.00, 1.50, and 2.00 if all resistors are of equal value. Note that R3 becomes the calibrating component of the network.

The input voltage to be compared is applied to pin 4, 6, 8, and 10 via collector Q1. Transistor Q1 conducts via TD (transducer) and R5. If no transducer is plugged into the socket, then Q1 is cut-off and the regulated voltage at the collector of 4.9 volts is applied to these inputs. This causes the comparator input to rise above the reference voltages and all LEDs come on.

The sensing element such as a thermistor (PTC type) was tried

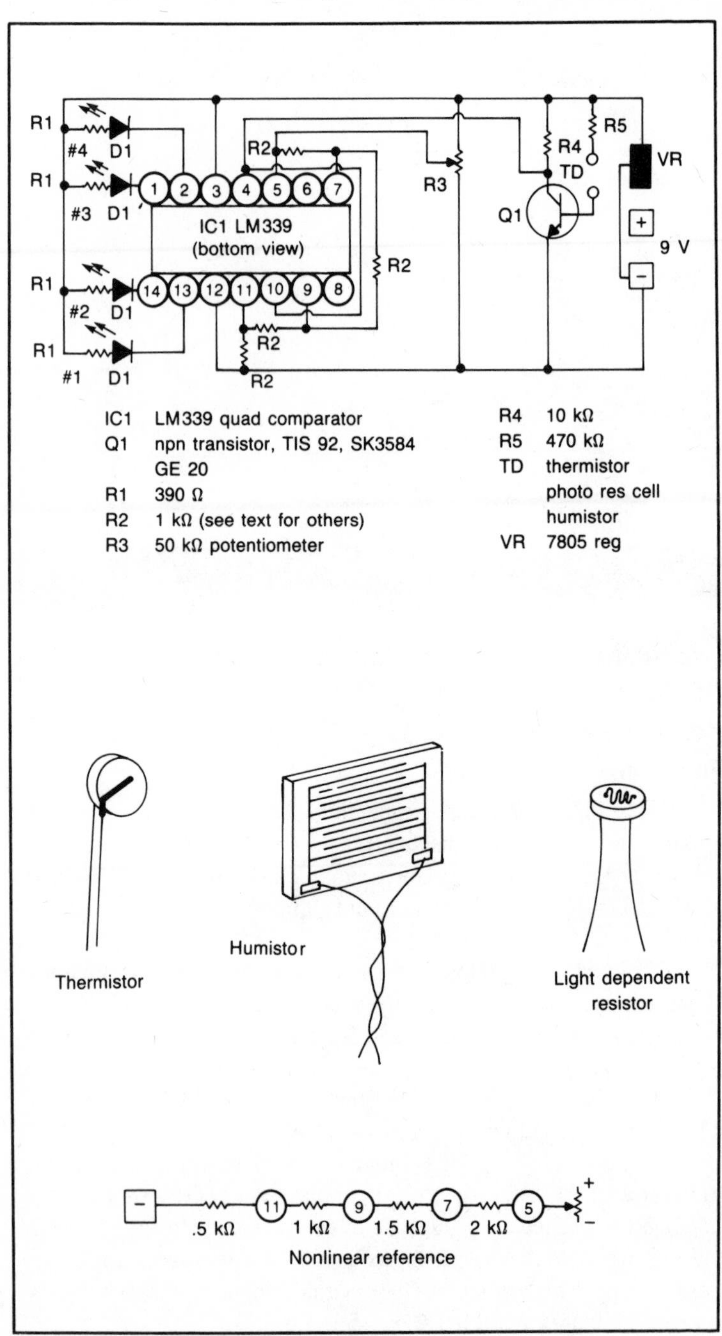

Fig. 8-14. Bar graph measurement (temperature, light, humidity).

in the circuit; it was 35 ohms at 20 degrees C (68 degrees F) and rose to above 200 kΩ at 100 degrees C (212 degrees F). Remember the conversion formula is $F = (1.8 \times C) + 32$. When plugged into the socket and temperature activated, the base circuit is complete via R4 and the supply voltage, thus Q1 is on. Q1 conducts as a result of this sensing and the voltage drop across R4 causes the voltages at the comparators to go below each reference voltage in sequential order from #1 to #4 LEDs. This condition makes all LEDs off.

When heat is applied to the thermistor, the resistance rises (remember it is a positive temperature coefficient type) to decrease the bias current and thus the collector current decreases. The voltage at the collector begins to rise. At specified voltages and equivalent temperatures, each LED comes on, and becomes a temperature reference. You would need to use a thermometer to adjust and calibrate the measurement. This unit would be excellent to measure the rise in water temperature as a way of checking the opening of an automotive thermostat.

The three point regulator (IC) is in the circuit to keep the unit accurate as the 9-volt battery begins to lose power.

Further Study

Try a light dependent resistor as the transducer to measure light levels. Make a humistor on an etched board as shown in Fig. 8-14. There should be enough moisture from your breath that blowing on it would turn on the LEDs. Remember you can use unequal resistors for R2 to provide unequal spacing and scales. (See nonlinear reference resistors in Fig. 8-14.) Can you connect a npn transistor and a triac to shut-down a circuit when one of the LEDs comes on?

PROJECT 52
DOT/BAR DISPLAY
(LIGHT, TEMPERATURE, HUMIDITY, AUDIO)

Function

The LM3914 dot/bar display driver is useful in many applications depending upon the creative thinking of the designer. This circuit will measure or monitor several quantities; however the explanation given will be on temperature with the thermistor as the sensor. (See Fig. 8-15.)

IC1, LM3914, has ten LED outputs in bar or dot turn-on as determined by S1. With S1 switching pin 9 to the positive source, the output is a bar display, whereas switching pin 9 to pin 11 will produce a dot sequence. This IC is a type of comparator in which a reference voltage of 1.2 volts is built into the circuit and extended out via pin 7 and controlled by pin 8. Since pin 6 is the high reference level, it is tied to pin 7, which is 1.2 volts. In this arrangement each LED output has a differential turn-on of .12 volt or (1.2/10). In other words, the first LED comes on at .12 volt and the last at 1.2 volts. By placing a 25 kΩ rheostat between pin 8 and ground, the reference voltage is adjustable from 1.2 to 2.7 volts. This provides for adjustment to set the LED turn-on with different transducers as a way of calibrating.

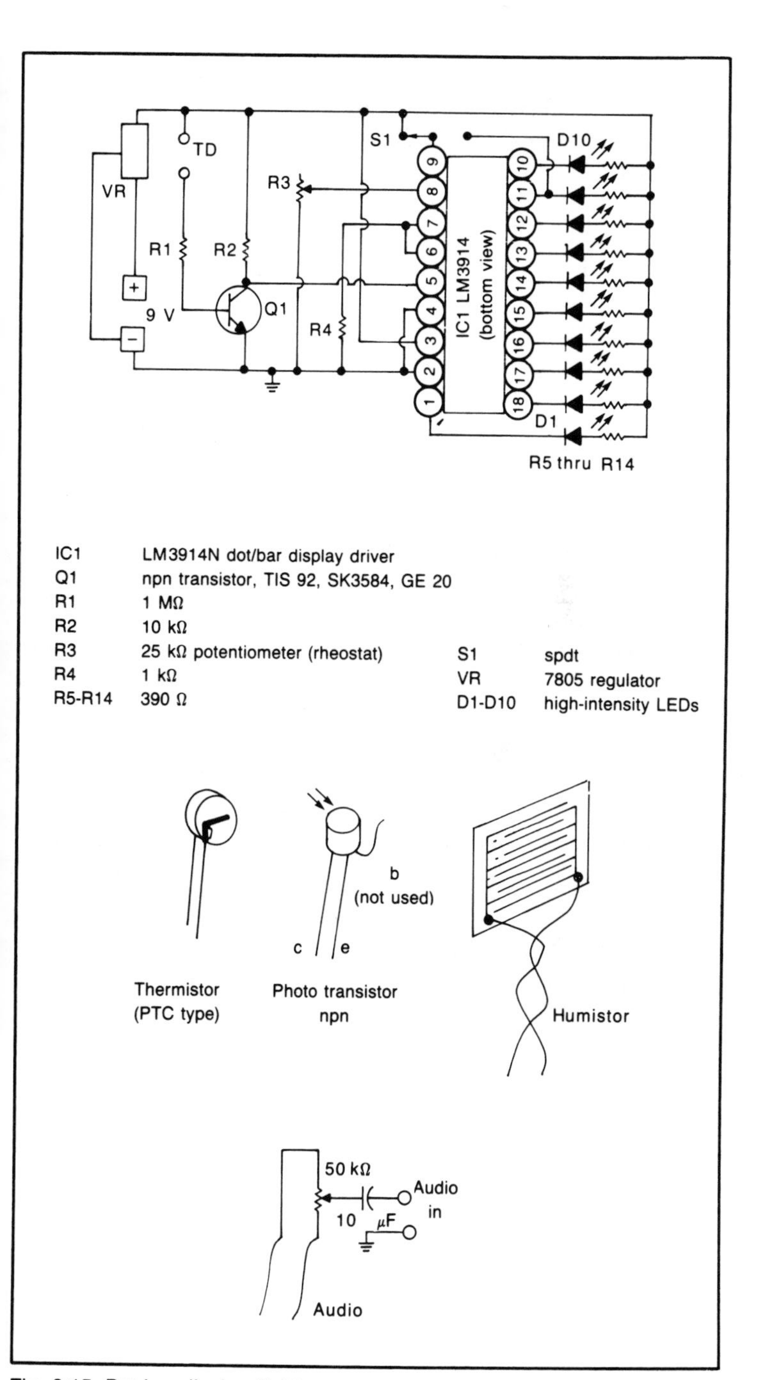

Fig. 8-15. Dot-bar display (light, temperature, humidity, and audio).

IC1 is interfaced with the sensor via Q1 for greater sensitivity and control. The bias is removed from Q1 base and the transistor is cut-off with no transducer in place. The voltage at the collector is the regulated supply voltage of 4.9 which makes the input pin 5 of IC1 above the 2.7 reference voltage set at pin 8 and all LEDs are on.

The thermistor sensor had a resistance of 35 ohms at ambient temperature. It is in series with bias transistor R1, therefore, Q1 is biased on and the voltage at the collector falls below 2.7 volts and all LEDs are off. If LED1 is on, you can adjust R3 reference set to just turn off as the starting point.

When heat strikes the thermistor, the resistance rises because it is a positive coefficient type. A portable hair dryer or heat gun can be used as the heat source. The resistance of the thermistor is proportional to the heat. As the thermistor resistance increases, the bias current decreases to cause the collector voltage to rise toward the supply voltage. When the voltage for each comparator is met, it turns on an LED in sequence for a bar display. If IC1 is in dot mode via S1, each LED will come on with the following-LED going out.

The humistor works in a reverse procedure. When the sensor is plugged into the socket, the circuit is still open and all LEDs are on. Just a slight blow of your breath will be sufficient to add humidity to the sensor to complete base circuit and the LEDs will go out from top toward bottom.

Further Study

Try a phototransistor or photodiode and determine the amount of light loss to turn on LEDs. Also, try an audio input with the circuit shown and the LEDs should blink; you will need to set the drive level with a 50 kΩ potentiometer (see Fig. 8-15). Can you connect a transistor and a triac at any pin to cause shutdown in bar display when that temperature has been reached? Can you add a SCR to self-latch at that temperature?

PROJECT 53
PREFERRED CHANNEL FLASHER (LIGHT RHYTHM)

Function

A preferred channel rhythm flasher can be simulated for channel marking and navigation. It would be useful as a teaching device for a boating course. It consists of two flashes, a single flash, followed by a delay off and then repeat. (See Fig. 8-16.)

IC3 consists of dual low-frequency oscillators that are activated via reset pins 4 and 10 respectively. When pins 4 and 10 are HIGH, the oscillators are on. Side 1 (pins 1 through 6) drives Q1 via pin 5 to flash L1. Side 2 (pins 8 through 13) drives Q2 via pin 9 to flash L2. The cycle of on/off/flash rate is controlled by IC2 for side 1 via pin 1. Side 2 is controlled via a monostable (one shot) circuit via pin 9. Note the monostable is triggered via IC2, pin 1, also.

When pin 1 of IC2 goes HIGH via a clock pulse from IC1, it toggles IC3, side 1 to drive Q1 for two flashes. IC2 toggles on the positive transition of the pulse, therefore it takes three pulses to complete one cycle.

The monostable IC1, side 2 triggers on the negative transition of the pulse from IC2, pin 1, thus it turns on IC3, side 2, when IC3 side 1 is off. The time constant for the monostable is of short duration to permit only one flash. The delay for the next full cycle is caused by a wait for another transition.

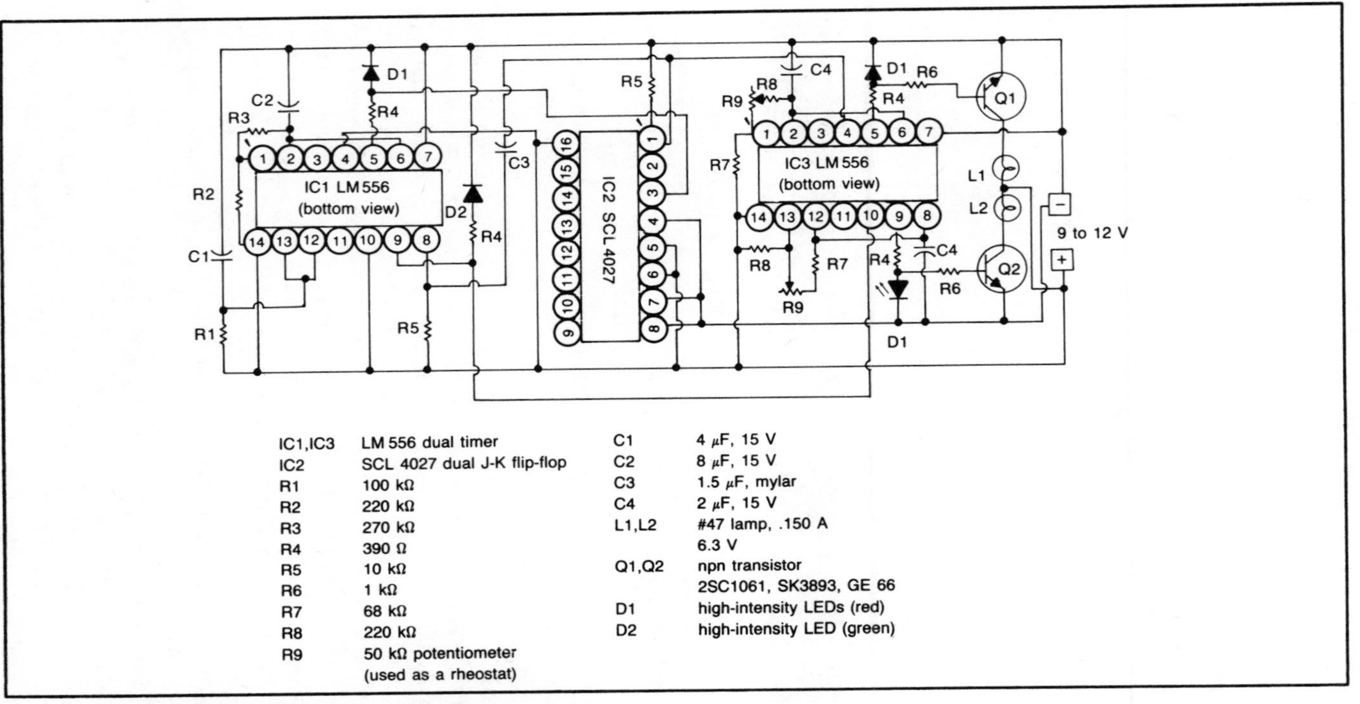

IC1,IC3	LM 556 dual timer
IC2	SCL 4027 dual J-K flip-flop
R1	100 kΩ
R2	220 kΩ
R3	270 kΩ
R4	390 Ω
R5	10 kΩ
R6	1 kΩ
R7	68 kΩ
R8	220 kΩ
R9	50 kΩ potentiometer (used as a rheostat)

C1	4 μF, 15 V
C2	8 μF, 15 V
C3	1.5 μF, mylar
C4	2 μF, 15 V
L1,L2	#47 lamp, .150 A 6.3 V
Q1,Q2	npn transistor 2SC1061, SK3893, GE 66
D1	high-intensity LEDs (red)
D2	high-intensity LED (green)

Fig. 8-16. Preferred channel flasher (light rhythm).

Trimmer resistors (R9s) are utilized to adjust the oscillator IC3 to provide some control on flash length.

Further Study

Other rhythm flashes are possible by changing the value of C1 of the monostable (try a 6 μF). If you make it too long, the flashing will extend over to occur the same time L1 is flashing.

PROJECT 54
S.O.S. LIGHT FLASHER

Function

This circuit simulates a continuously flashing SOS signal that could be valuable in an emergency. It would be a beacon for sight location in addition to signaling an emergency condition.

IC1, is an astable multivibrator that operates at a rate of approximately 1 pulse per second or 30 hertz per minute as determined by R1, R2, and C1. The output is monitored by D1 and is coupled to IC2, JK flip/flop pin 3, as clock input. (See Fig. 8-17.)

One of the two JK flip/flops is utilized in IC2. IC2 is connected to toggle the output at pins 1 and 2 with every positive transition of the clock. The outputs are connected to the reset pins 4 and 10 of IC3 in order to turn each one on alternately.

Side 1, pins 1 through 6, of IC3 is the faster running multivibrator as determined by R7, R8, R6, and C3. Side 2, pins 8 through 13 is the slower multivibrator as determined by R4, R5, R8, and C4. The outputs at pins 5 and 9 go to drive Q1 and Q2 via R11 and R12 respectively. The transistors drive the #47 lamps to produce flashing lights.

IC1 determines the length of time on and thus the number of flashes for each lamp. The components shown in the schematic provide three short flashes, three longer flashes, and three short flashes

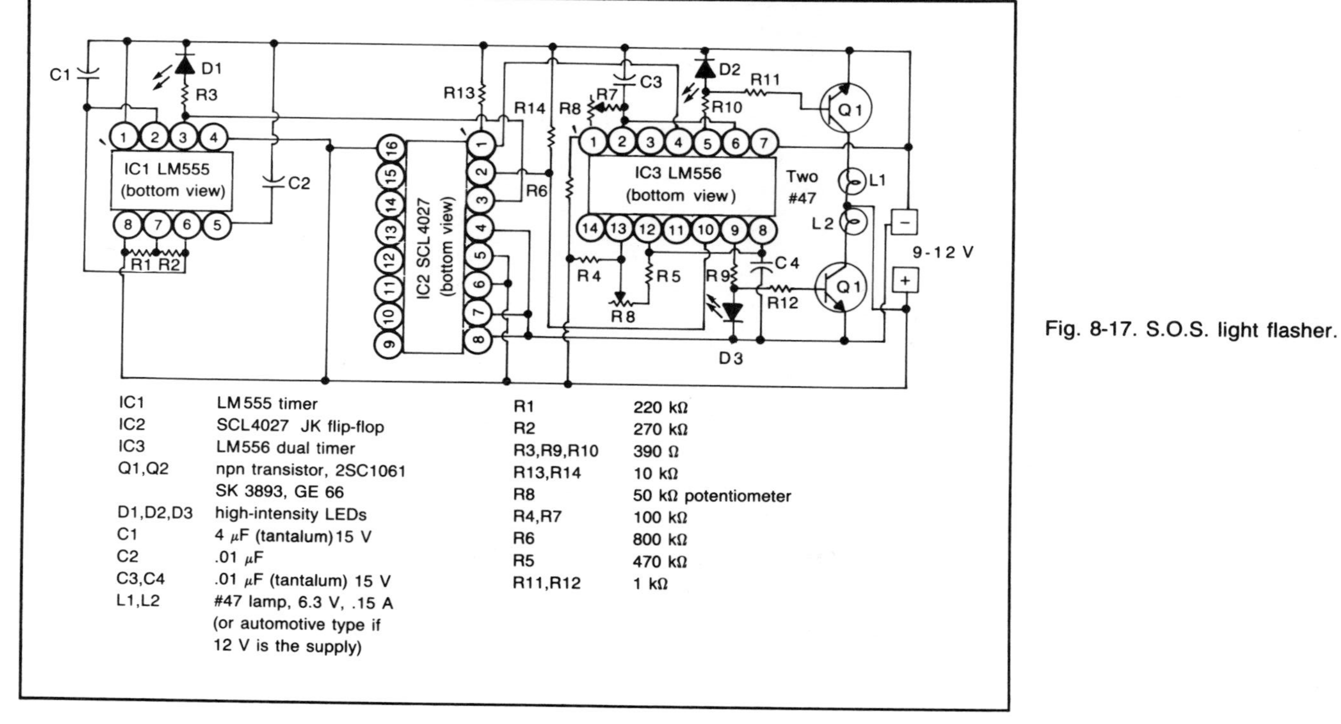

Fig. 8-17. S.O.S. light flasher.

to complete the S.O.S. cycle. The potentiometers (R8s) are in the circuit to trim the time cycle for each multivibrator. The components are critical in the timers; be sure to use tantalum capacitors.

Further Study

Change the value of C1 to 8 μF and note the change in flash rate. Other rhythmic flash rates are possible by changing C3 and C4—go ahead and try. It is possible to connect one reset directly to the IC1 output and get concurring flashes of the two lamps.

PROJECT 55
TURN SIGNAL FLASHER
(WITH 4-WAY SAFETY FLASHER)

Function

A turn signal flasher could be valuable for a bicycle or motorcycle. It could be an add-on for a small trailer or for a towed vehicle. In addition it has an override switch to make the unit into a safety hazard signal. (See Fig. 8-18.)

IC1, LM556, is a dual timer in astable multivibrator mode that oscillates at the frequency of 1 hertz per second as determined by C1, R1, and R2. The outputs at pins 5 and 9 are coupled to Q1 and Q2 via bias resistors (R1s). Right and left lamps are switched on at each pulse when oscillation takes place.

The right and left turn-on of the multivibrators is accomplished through S1; a single-pole double-throw (normally open) spring-loaded switch. This switch activates each multivibrator by switching resets high; both reset pins 4 and 10 are held low via R5s until switching takes place.

The dashed lines on the schematic indicate the added circuit in the event you wish both front and rear signals. This may be omitted if you want only rear signaling.

Switch S2 provides a way of switching on both multivibrators at the same time, and thus is a safety hazard feature.

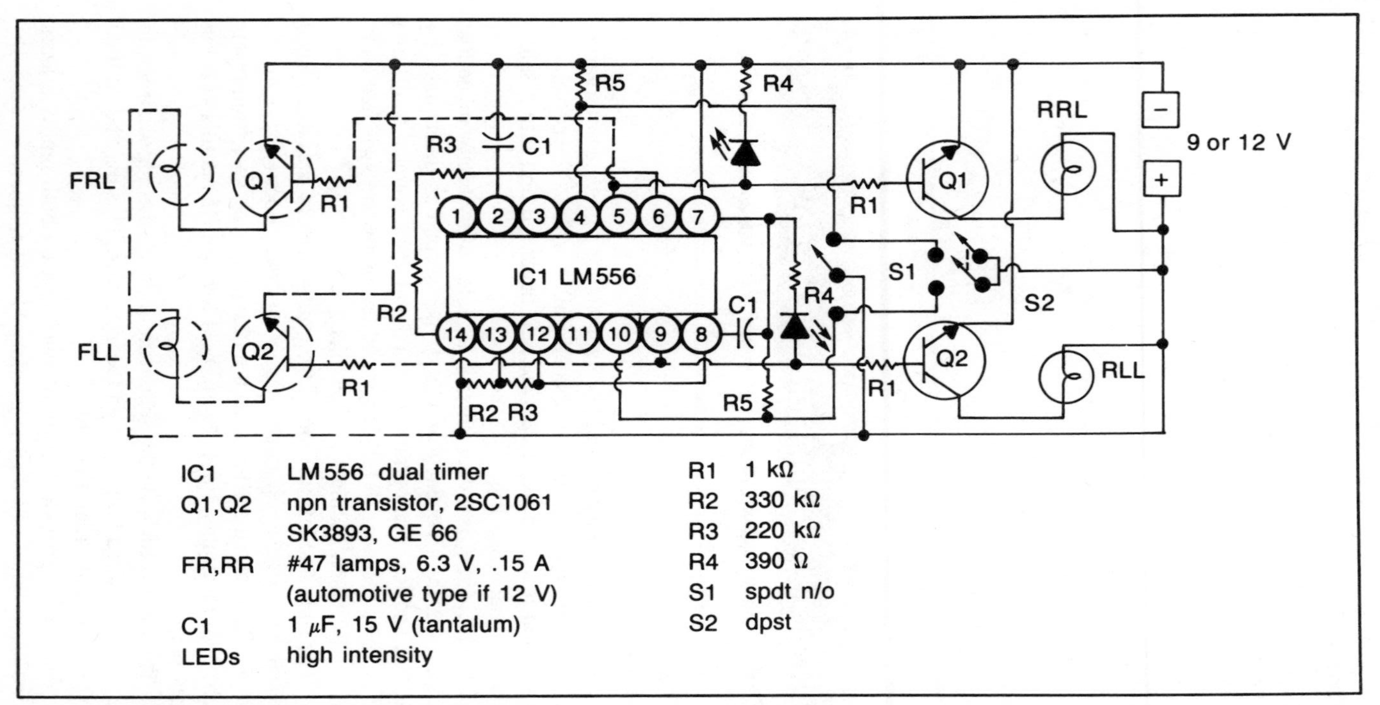

Fig. 8-18. Turn signal flasher (with 4-way safety flasher).

Further Study

Change the value of C1 in both multivibrators to increase or decrease the flash rate. The LEDs are in the circuit to monitor the outputs from the multivibrator in case of lamp failure.

PROJECT 56
THREE-STATE ALARM

Function

The three-state alarm (flashing lights) will flash two LEDs when all three conditions at the input have been met. This would be useful in monitoring a situation as to the level of light, liquid, and humidity, or when any three switches have been closed. (See Fig. 8-19.)

One third of IC1, 3-input AND gate, will go HIGH at the output pin 6 if all three inputs go HIGH. The input sensors could be three electromechanical switches with trip cords. When all three trip cords are broken, the inputs at pins 3, 4, and 5 go HIGH as a result of TR1, TR2, and TR3 being closed.

The output pin 6 is connected to the reset pin 4 of IC2. When pin 4 of IC2 goes HIGH, oscillation occurs to flash the LEDs. When pin 4 is LOW, the output pin 3 of IC2 is LOW and LED D3 (green) is on to mean everything is OK. When the three-state conditions are met, the oscillator flashes the red and green LEDs alternately.

A thermistor (8 megohms-NTC type) was utilized with two phototransistors. The negative pull-down resistor was changed from 10 kΩ to 470 kΩ at pin 3 and a thermistor was used in TR1. The LEDs flashed when light was applied to the phototransistors and heat to the thermistor. The thermistor and the negative pull-down resistor should

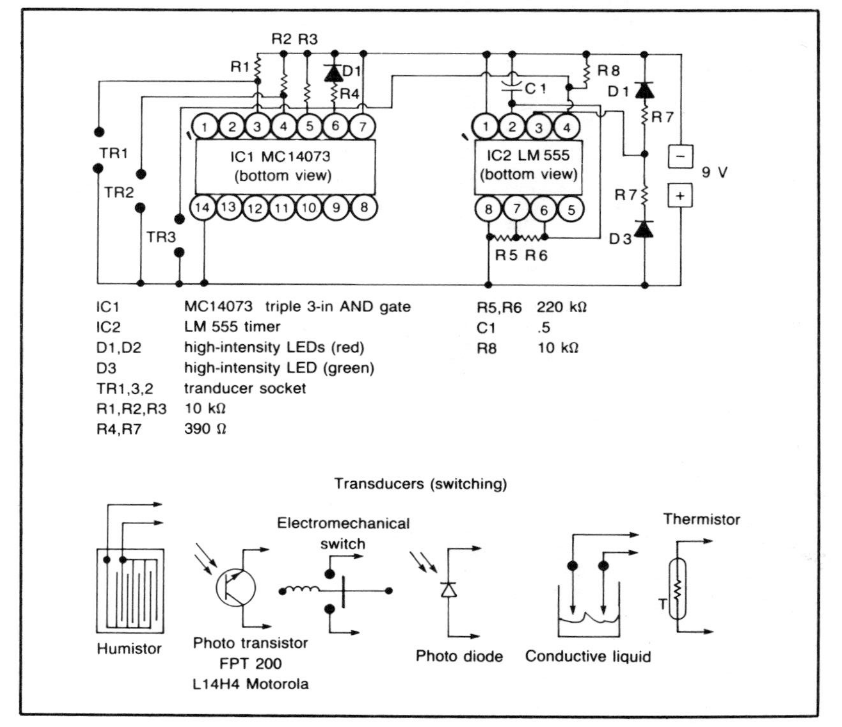

Fig. 8-19. Three-state alarm.

have a ratio of 16 to 1 at ambient temperatures to work effectively. The lowest ratio that can be utilized is 3 to 1 at ambient temperatures. A reduced ratio will mean less temperature change to meet the high condition.

Further Study

Try other transducers to activate the circuit. (See Fig. 8-19 for other transducers.) Remember you can tie in any single-input high and thus activate the circuit with only a two-state condition.

PROJECT 57
FOUR-STATE ALARM
(WITH LATCH AND POSITION MONITOR)

Function

The four state alarm/monitor would be useful to detect intrusion at four locations. In addition, it tells via LEDs the exact location where the intrusion took place. (See Fig. 8-20.)

One half of IC1, MC14072, 4-input OR gate, pins 1 through 5 is used to monitor four states or conditions. The output pin 1 goes high if any one of the four inputs go high. The four detectors connected to the TR1 through TR4 sockets must be of a resistance less than 1 kΩ to hold each input low and thus the output low. When any one transducer (switch) is opened or resistance rises above 1 kΩ, the output at pin 1 goes high.

Since pin 1 is connected to the gate of the SCR via R4, the SCR turns on to produce a rising voltage at the cathode. The cathode end of the SCR is connected to pin 4 of IC2, therefore, the reset of IC2 is HIGH and the oscillator comes on to flash the LEDs alternately. This signals that one of the transducers (detectors) has been activated.

The SCR in dc operation latches if the current through the SCR is above the holding current. Resistor R3 is the only load on the SCR and its resistance is low enough to produce about 3 mA, which is

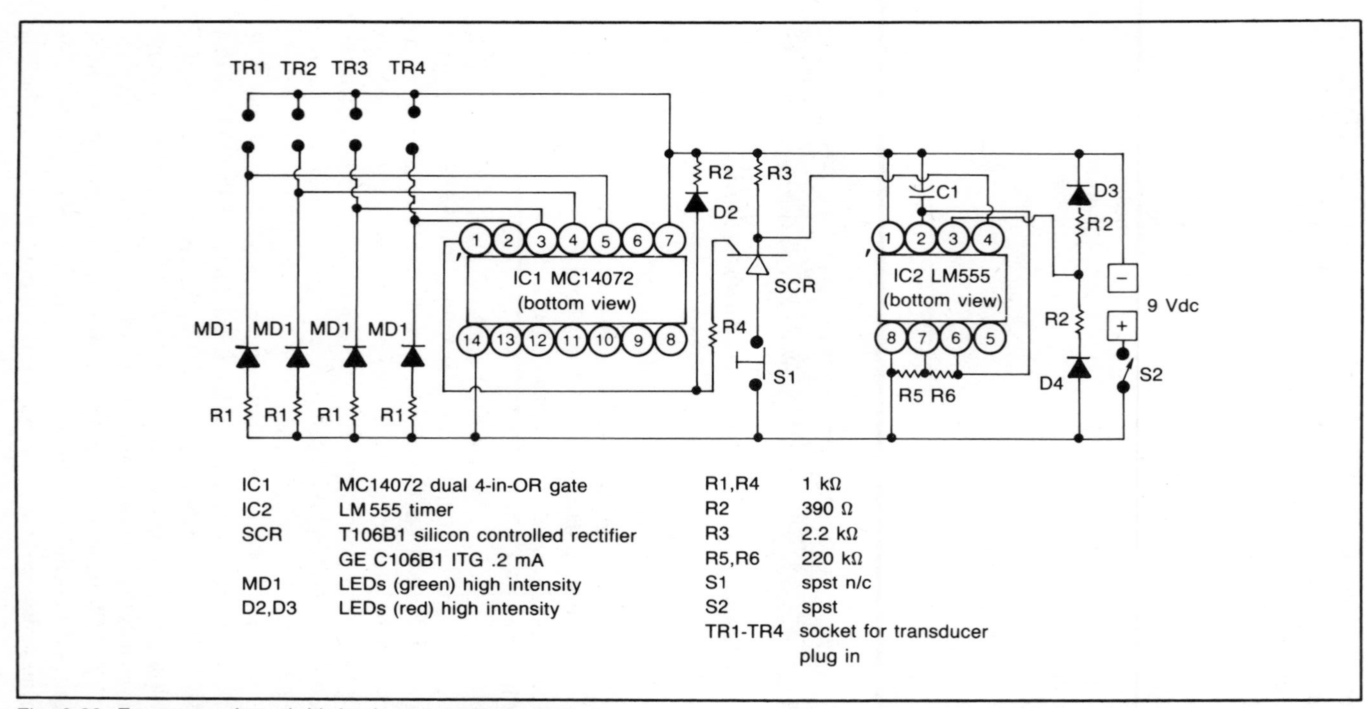

Fig. 8-20. Four-state alarm (with latch and position monitor).

above the holding current for SCR C106B1. Switch S1 is in the circuit to commutate the holding current and thus reset the SCR after the detector control IC1 is reset at the inputs. All the MDs (monitor detectors) at the inputs are on when IC1 has been reset.

The LEDs are in the circuit to monitor each of the input detectors. They are on when all detectors (sensors) are closed at TR locations. If a TR is opened or rises to a high resistance by environmental change, the LED (MD) goes out to tell at what location an intrusion has taken place. Resetting any TR cannot shut-down the flasher because it has latched via the SCR. The circuit is rearmed by resetting the TRs and pushing the switch S1 to reset the SCR, which in turn resets the flasher. The alarm could be an audible tone by redesigning the flasher circuit.

Further Study

Try a magnetic reed switch as a transducer. Can you change IC2 to produce an audible output in place of the flashing LEDs? Look elsewhere in this text for the audible oscillator from a LM555 circuit. Can you use a light dependent resistor as a transducer that will trigger an alarm if a light is turned off?

PROJECT 58
FLUID LEVEL INDICATOR

Function

The fluid level indicator warns of low level liquid in the container. It could be useful to notify someone when the windshield washer fluid is low or when the cistern is low and water conservation practice must take place. (See Fig. 8-21.)

The internal oscillator frequency of IC1 is 2-hertz per second that is riding on a pedestal of 8 volts. This frequency is coupled to the detector input pin 9 via C2, therefore this frequency drives the output load. It is possible to drive a #47 lamp or LED directly as shown in dotted line on the schematic and omitting the transistor circuit at the point X cut. Both the lamp and the transistor-LED circuits were operated simultaneously for experimental purposes.

When the probe is in a high resistance state (air) the output load flashes. When the probe is in a conductive liquid, the internal detector of IC1 is grounded and the oscillator frequency cannot reach the output; however the fluid in the container must be grounded as well as the minus supply. What one sees is a probe in the fluid that causes the internal detector to be grounded and the light is off. When the fluid is consumed and level falls below the probe, the light flashes indicating a low level in the reservoir. Adjusting the probe height

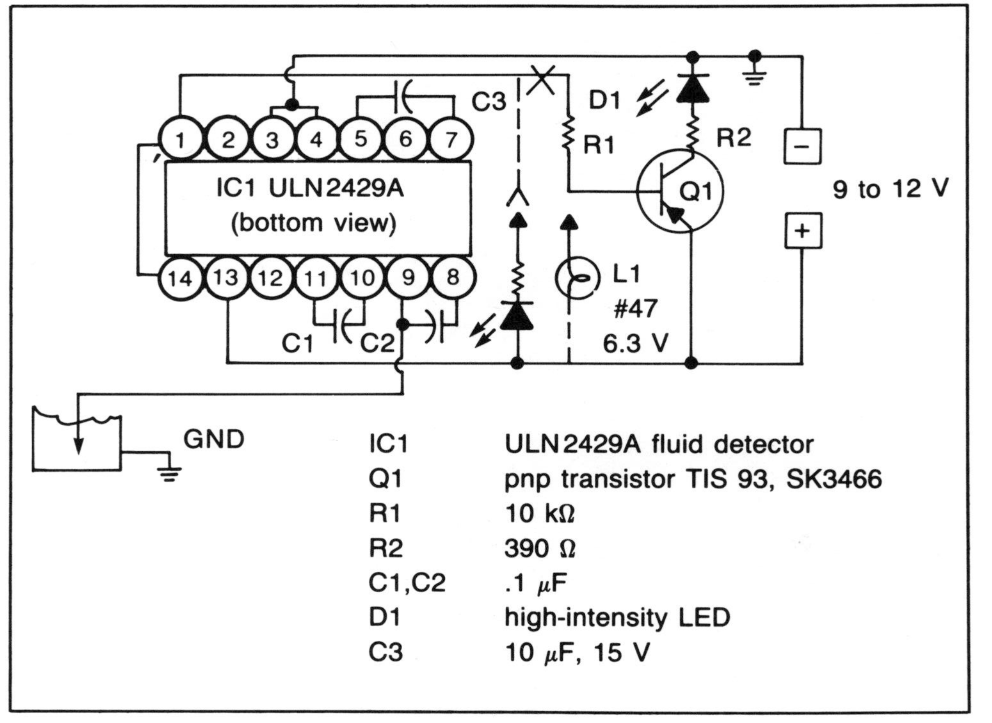

Fig. 8-21. Fluid level indicator.

from the bottom sets the level of detection. Decoupling via C1 is not a concern in ac mode and that component can be removed.

Further Study

Change the value of C3 to 5 μF and note the change in flash rate. Try different liquids like windshield washer fluid, water, juice, etc. (Avoid flammable liquids for safety reasons.)

PROJECT 59
HOUSE TEMPERATURE REMINDER

Function

One of the concerns when a family goes on vacation for an extended period of time in the winter is the heating condition inside the house. The concern is that the heating system is holding the heat so the pipes don't freeze or that the furnace is not overheating. A neighbor usually makes frequent checks on the condition. The circuit shown here is a house temperature reminder that can be viewed by looking through a window at the indicator rather than by physical entry. (See Fig. 8-22.)

ICs 1, 2, and 3 are LM741 operational amplifiers in comparative function. IC1 has a reference voltage of 4.5 set at pin 3 via R2 and R3. Thermistor 1, and R1 are the controlling components and R1 adjusts the voltage at pin 2 to be above the reference voltage. Output pin 6 is low to turn on the green LED. If the temperature in the house is maintained at middle limit (threshold), the green LED means the system is OK. However, if the temperature becomes excessive, thermistor 1 (NTC type) will fall in resistance due to the higher temperature. This causes the voltage at pin 2 to fall below preset voltage at pin 3 and output pin 6 goes HIGH to turn off the green LED.

IC2 gets its positive reference voltage at pin 3 via R5 and R6,

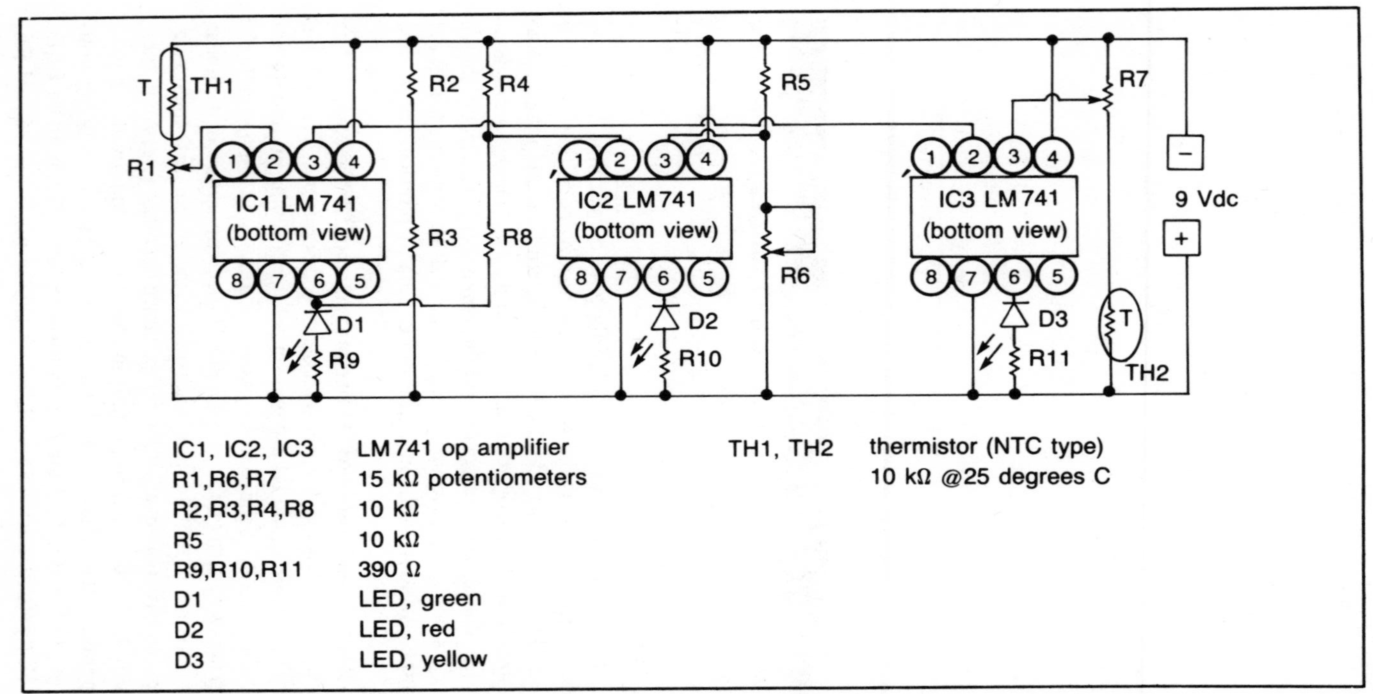

Fig. 8-22. House temperature reminder.

which is set just slightly below the voltage that will appear at pin 2. The change voltage at pin 2 will be determined by R4, R8, and the positive voltage that will come on at output pin 6 when it goes high. The result of this action will be that the red LED at pin 6 of IC2 will turn on when the green LED goes off to remind one that the heat is too high.

IC3, pin 2 is referenced at the same voltage at pin 3 of IC1 or 4.5 volts. The change voltage will appear at pin 3. Output pin 6 of IC3 is HIGH and the yellow LED is off because R7 sets the change voltage above the reference voltage. Thermistor 2 is at the top of the positive side of the change side and it increases in resistance with lowering of temperature, therefore the change voltage at pin 3 will fall. If cold temperature is applied to thermistor 2, (NTC type), it will rise in resistance and because it is at the positive end, the voltage at pin 3 will fall. When the voltage at pin 3 falls below the reference voltage, the output pin 6 will go low to turn on the yellow LED.

What one sees is a middle limit with IC1 and a green LED, an upper limit with IC2 and a red LED, and a lower limit with IC3 and a yellow LED. Note both green and yellow LEDs will be on at the same time.

Further Study

Complete the following measurements to understand action:

IC1 Circuit:

1. Measure reference voltage at pin 3.
2. Measure the change in voltage at pin 2.
3. Adjust R4 while observing voltage at pin 2 and the state of the green LED.

IC2 Circuit:

1. Measure reference voltage at pin 3. Adjust with R6.
2. Measure change voltage at pin 2.
3. Adjust R1 to make the green LED go out while observing the voltage at pin 2.
4. Can you add another thermistor in the pin 3 reference voltage circuit to delay the red LED to turn on at a higher temperature than the green LED control?

IC3 Circuit:

1. Measure the reference voltage at pin 2. It should be the same as pin 3 of IC1.
2. Measure the change in voltage at pin 3 and vary R7.

Finally, test the circuit by heating a paper towel and covering TH1. Cool a piece of paper towel in the refrigerator or freeze and cover TH2. Get a thermometer and calibrate the unit.

PROJECT 60
ELECTRONIC FLASH/STROBE

Function

The electronic flash can be utilized in photography for direct light source or for fill light for special effect or shadow cancel. The unit shown here could be useful with the photoflash slave circuit in this book. It can be used with slight modification as a slow-pulse strobe light for boating or another recreational safety device. (See Fig. 8-23.)

IC1, LM386, is a low-voltage audio amplifier. The IC is in an astable multivibrator mode with a frequency of 33 kilohertz as determined by C1 and R3. The output at pin 5 is a square wave that drives T1 via C2.

T1 is a 120 V 60 hertz, step-down transformer connected in reverse direction with the normal 6.3 V secondary as the primary. The P-P voltage ac at the secondary is about 400 that delivers 200 volts dc conversion by D1 to C3 and C4. Use caution in working with the circuit because this kind of voltage can deliver a heavy sting. Remember if the capacitors C3, C4, or C5 are removed from the proto-board, they should be discharged before handling. The dc voltage is applied to C5 via R4 for trigger potential. The charge time for full voltage to C5 is about .5 of a second which means rapid recycle.

A Xenon flash tube requires a high trigger voltage to fire, which

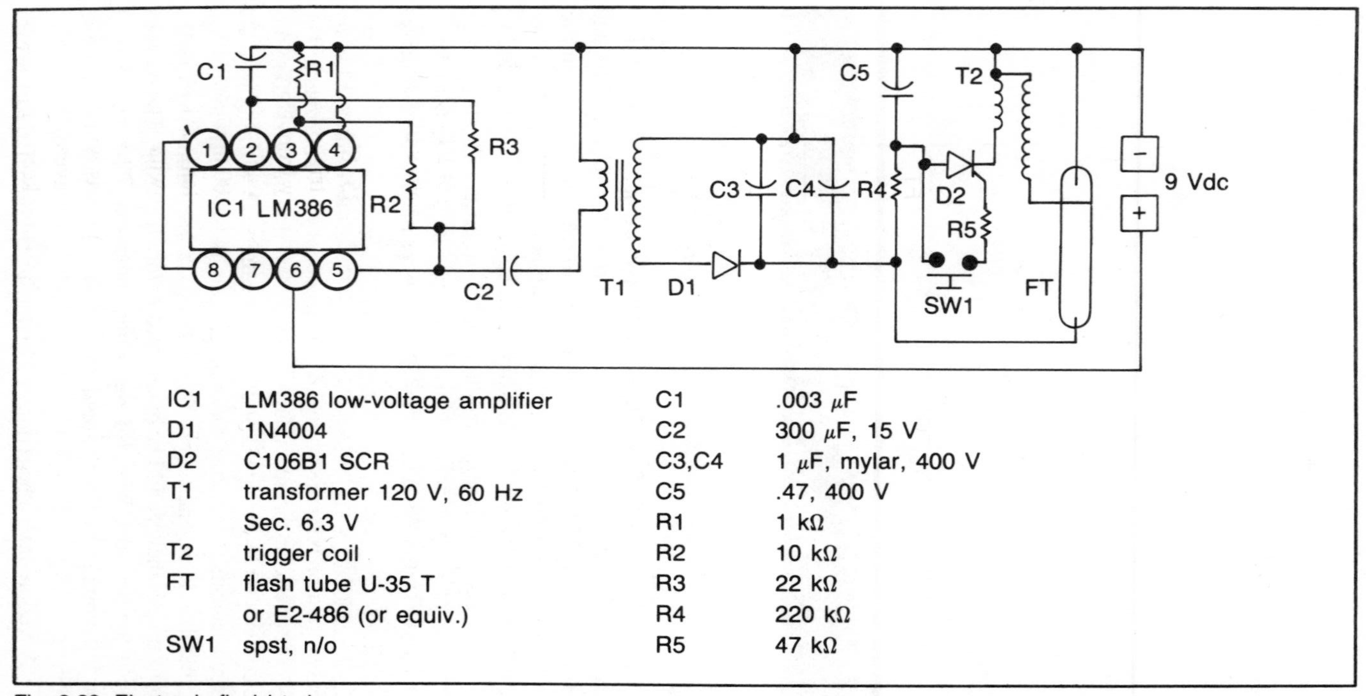

IC1	LM386 low-voltage amplifier	C1	.003 μF
D1	1N4004	C2	300 μF, 15 V
D2	C106B1 SCR	C3,C4	1 μF, mylar, 400 V
T1	transformer 120 V, 60 Hz	C5	.47, 400 V
	Sec. 6.3 V	R1	1 kΩ
T2	trigger coil	R2	10 kΩ
FT	flash tube U-35 T	R3	22 kΩ
	or E2-486 (or equiv.)	R4	220 kΩ
SW1	spst, n/o	R5	47 kΩ

Fig. 8-23. Electronic flash/strobe.

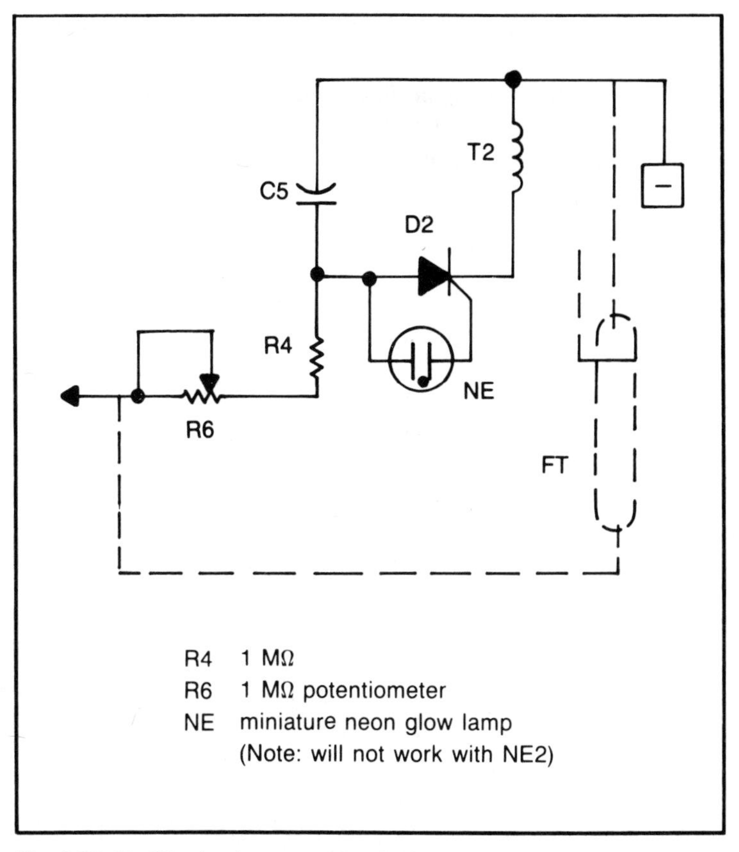

Fig. 8-24. Modification for strobe/slow flash rate.

is delivered by the trigger coil to the trigger anode of FT. When FT is triggered, C3 and C4 which are charged to 200 volts will discharge that energy to produce the high-intensity flash.

The trigger circuit consists of C5, C4, D2, R5, SW1 and T2. The SCR, D2, serves as the transfer switch to the trigger coil. The switching of the 200 volts to the trigger coil primary from C5 would produce a heavy arc that would be detrimental to any mechanical switch over time. The SCR triggers with as little as 200 microamperes in the gate, thus resistor R5 will reduce this problem. SW1 in the circuit is the manual switch. It would be replaced with the camera synchronization input or slave switch.

The standby current when charge is complete is about 30 mA while the operational current during recharge will average 60 mA; however, the recycle time is less than 1 second to keep battery life at an acceptable limit. A neon ready light with a resistor could be placed across C4, but it was not in this design because the recycle is short compared to similar flash units. Inexpensive flash tube assemblies with tube, reflector, lens, and leads that are ideal for this application are available from several electronics supply houses.

Further Study

The circuit can be modified as shown in Fig. 8-24 to produce a slow-pulse flash strobe like 1 or 2 flashes per second. A higher flash rate is difficult in this design because the high voltage becomes below flash-tube fire level in a rapid recycle requirement. One might experiment with a 120:4 transformer for a higher output voltage to produce a higher intense flash and/or strobe rate possibility.

PROJECT 61
PHOTOFLASH SLAVE

Function

A photoflash slave is useful to the photographer for background fill, highlight effect, or increased light intensity for large spaces or distances.

IC1, CD4071, a quad two-input OR gate is utilized as a trigger driver to an SCR (switch) to turn on a reed switch that in turn fires an electronic flash. The output and transition time of an OR gate at 10 volts is 10 to 15 nanoseconds which switches instantaneously and within photographic flash capabilities. (See Fig. 8-25.)

The truth table of an OR gate states that if either or both inputs go HIGH, the output goes HIGH. Pin 4 of gate 1 is held LOW to the B−. Pin 2 goes to the emitter of phototransistor Q1 and is held LOW via R3 and Q1 in the cutoff state. If pin 2 goes HIGH as the result of Q1 conducting, output pin 3 goes HIGH. Pins 6 and 5 of gate 2 are connected to pin 3, therefore they go HIGH. Both pins HIGH cause output pin 4 to go HIGH. The SCR (as a dc switch) turns on because the gate is connected to pin 4 via R7.

When the SCR is gated on, it conducts to turn on LED D1 and S3. LED D1 is in the circuit to monitor and verify the application of energy to the reed relay S3. The magnetic field coil of S3 closes

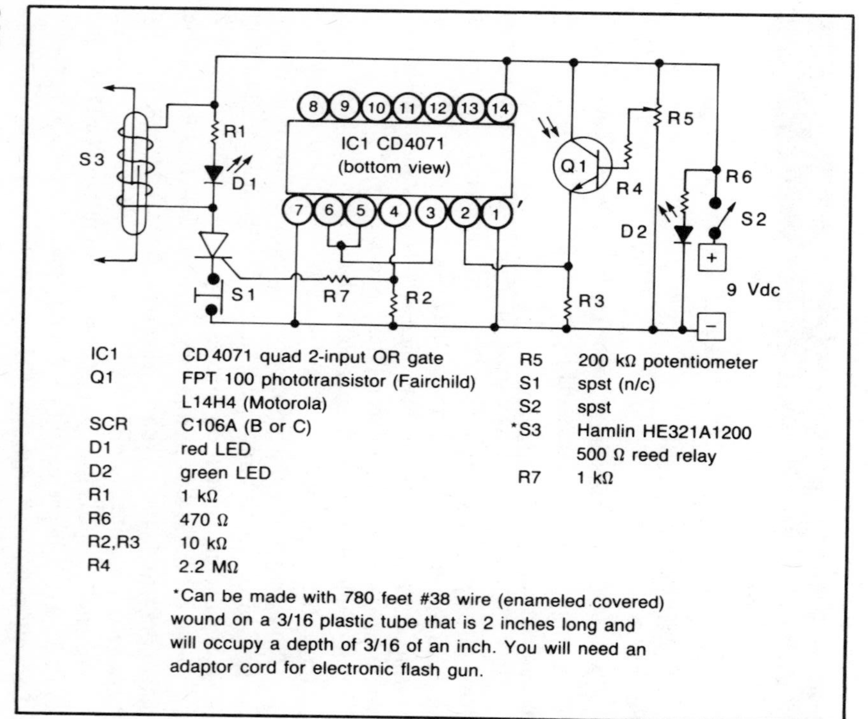

IC1	CD 4071 quad 2-input OR gate	R5	200 kΩ potentiometer
Q1	FPT 100 phototransistor (Fairchild)	S1	spst (n/c)
	L14H4 (Motorola)	S2	spst
SCR	C106A (B or C)	*S3	Hamlin HE321A1200
D1	red LED		500 Ω reed relay
D2	green LED	R7	1 kΩ
R1	1 kΩ		
R6	470 Ω		
R2,R3	10 kΩ		
R4	2.2 MΩ		

*Can be made with 780 feet #38 wire (enameled covered) wound on a 3/16 plastic tube that is 2 inches long and will occupy a depth of 3/16 of an inch. You will need an adaptor cord for electronic flash gun.

Fig. 8-25. Photoflash slave.

the reed switch to fire the flash. An SCR in dc mode latches when energized even if the gate energy is removed; S1 is placed into the circuit as a reset to turn off the SCR after each flash.

The major control begins with Q1, a phototransistor that can be biased on with light. In addition, base bias is in the circuit to increase sensitivity and provide additional control over the turn-on of Q1. Q1 can be made sensitive to turn on with a flashlight. The sensitivity control, R5, should be adjusted to prevent triggering by room lighting. Trial runs demonstrated that direct flash as far as 40 feet would trigger the slave flash. Flash bounced off a 12-foot ceiling had no difficulty in triggering the unit.

A shroud to go over the transistor was constructed of a black alligator-clip sleeve. This is necessary to give directional control and prevent incident light from causing false triggering.

Further Study

Try further sensitivity control by adding a rheostat in series with R7. Try different settings of R5 with several light sources and/or flash distances to check sensitivity.

PROJECT 62
HEADLIGHTS ON REMINDER

Function

Many automobile drivers have experienced the embarrassment of having left their parking lights or headlights on after departing, only to find their car unstartable upon return due to battery drain. The circuit shown here reminds the driver that the headlights are on when turning off the ignition. It can be used for a rear view defroster or any other device that has a heavy drain on the battery. (See Fig. 8-26.)

IC1, CD4011, a quad 2-input NAND gate has two gates cross-coupled as a slow-pulse multivibrator as determined by R2, R3, and C1. The output at pin 4 is coupled to input pins 8 and 9 of a third gate via C2. Every time 8 and 9 go HIGH the output pin 10 goes LOW to turn on the LED, D1. The flash rate is about 8 pulses per second.

Pin 1 of the cross-coupled gate must be HIGH to enable oscillation. Pin 1 is connected to output pin 11 of the fourth gate. Input pin 12 is HIGH to the B+ through the headlight switch along with power-up pin 14. Input pin 13 is held HIGH via the ignition switch. In this state output pin 11 is LOW with the ignition switch on and the cross-coupled oscillator is disabled.

If the driver turns off the ignition key, the B+ voltage is removed

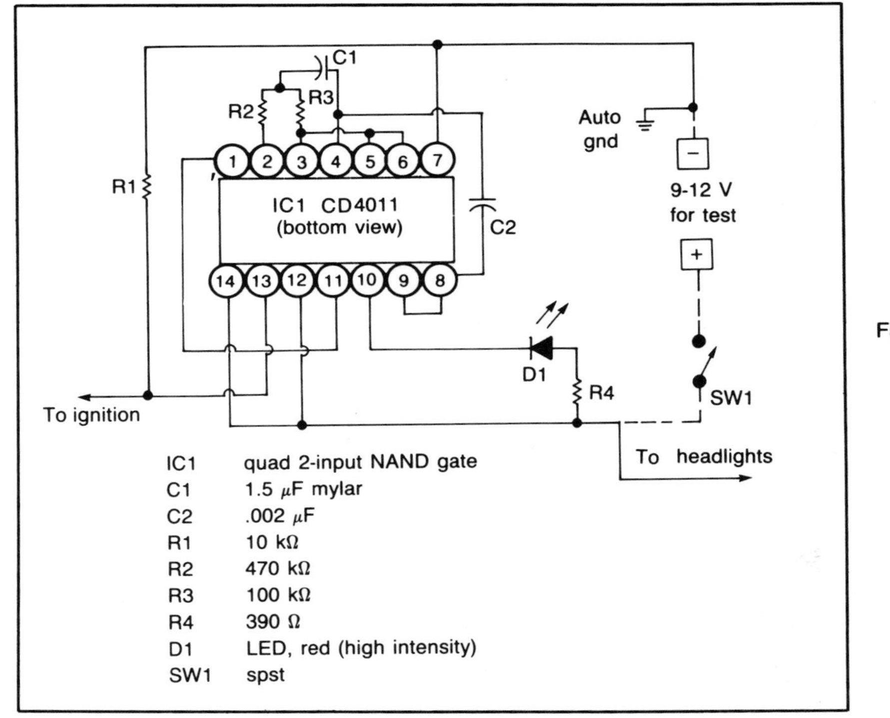

Fig. 8-26. Headlights on reminder.

from pin 13 of the fourth gate and it goes LOW via R1. One input LOW with the other input HIGH with the lights on will enable the oscillator to flash a warning to the driver.

If the light switch is off, the oscillator is disabled via loss of power-up voltage to pin 14. The condition of lights off means no warning even with the ignition on. Yet with lights on and ignition on, all is satisfactory and no warning. But if lights are on and ignition is turned off, the flashing LED reminds the driver that the lights are on.

The dashed lines on the schematic show a substitute voltage arrangement for testing the circuit prior to installation in the automobile or for laboratory analysis of a simulated proto-board layout.

Further Study

You can change the value of C1 to change the flash rate. You can add a seven-segment readout to flash a letter L for (lights). (It must be common cathode to pin 10 and the other elements to B+ with an appropriate resistor.)

PROJECT 63
DISTURBANCE MONITOR WITH LATCH

Function

The disturbance monitor is useful to record the rearrangement of papers on a desk, drawer disturbance, or articles moved. It can be built as a paperweight that in time-delay-on will monitor a change and it cannot be stopped by resetting the trigger mechanism. (See Fig. 8-27.)

One side of IC1, AND gate, is the monitor and latch circuit. The AND gate has a truth table that states when both inputs are HIGH, the output is HIGH. Pin 6 of gate 2 is HIGH and pin 5 is HIGH via R3 and the transistor Q1 base-to-emitter connection to B+. Therefore, this gate and the output pin 4 goes HIGH. Since output pin 4 of gate 2 is connected to pin 2 of gate 1, it is held HIGH. Pin 1 of gate 1 is held HIGH via jumper lead across C1. All inputs are HIGH and the unit is set.

Transistor Q1 with base positive is reversed biased and is in cut-off state. Monitor LED is out with Q1 turned off. If the trigger wire is moved to open the B+, capacitor C1 begins to charge via R1. The voltage across R1 is HIGH at the beginning of the charge time of C1; as C1 becomes charged, pin 1 goes LOW, thus output pin 3 goes LOW. Since output pin 3 of gate 1 is tied to input pin 5 of gate 2, gate 2

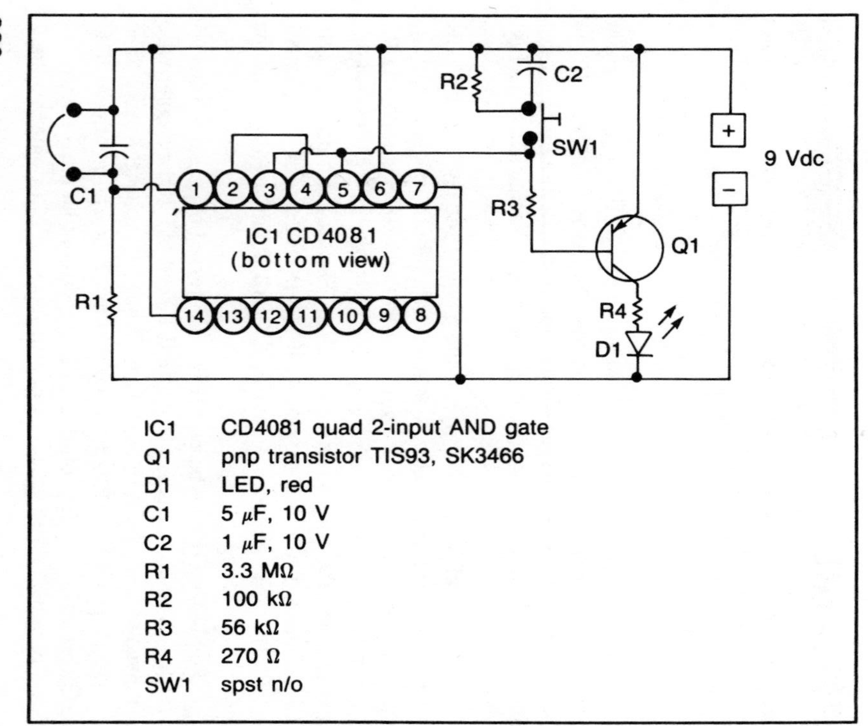

IC1	CD4081 quad 2-input AND gate
Q1	pnp transistor TIS93, SK3466
D1	LED, red
C1	5 μF, 10 V
C2	1 μF, 10 V
R1	3.3 MΩ
R2	100 kΩ
R3	56 kΩ
R4	270 Ω
SW1	spst n/o

Fig. 8-27. Disturbance monitor/with latch.

output pin 4 goes LOW. It is connected to pin 2 of gate 1, therefore, this gate cannot be toggled if input jumper is reestablished and we have a latched condition. When pin 3 went LOW it forward biased Q1 to turn on the LED.

S1 is in the circuit to reset Q1 to off by momentarily reversing the bias voltage to positive via C2. R2 is in the circuit to discharge C2 when S1 is released, thus getting ready for next reset. You cannot reset Q1 unless the jumper is reconnected across C1.

The delay turn-on is determined by R1 and C11. Making either one larger in value will increase the delay time. The delay time with the values shown is about 15 seconds. The collector current is 20 mA and the base bias current of Q1 is .11 mA in forward bias state.

Further Study

Change C2 to 10 μF and time the delay. You should be able to carry this design further to set off an audible alarm.

PROJECT 64
LIGHT BEAM TRANSMITTER

Function

A light-beam transmitter is a way of communicating by light wave on a line of sight. The light-beam modulation principle is found in motion picture projector sound systems in which the sound appears as a track on the side of the film. (See Fig. 8-28.)

IC1, LF351, is an operational amplifier with JFET input for high impedance. The input signal is applied to inverting input pin 2 and the amplification appears at pin 6 to load R5, which is a potentiometer for drive voltage control. The gain is a factor ratio of R2 to R3, thus with 1000/1,000,000 ohm values, the gain would be 1000. This is adequate for a low-level crystal microphone with 25 to 50 mV to drive the LM384 power amplifier. The LF351 requires two 9-volt batteries or a bipolar supply. IC3, the 7808 regulator supplies the positive input at pin 7 of IC1. A 9-volt battery at pin 4 supplies the negative voltage. An op amplifier requires a balancing system to counteract the offset voltage at pin 6. This is accomplished by measuring the voltage between pin 6 and common with a DVM and then adjusting R3 to get a zero or near zero reading (this is called null balancing).

IC2, LM384 is a straightforward power amplifier of 5 watts. Maximum voltage is 26 V and the gain factor is 50. The output must be

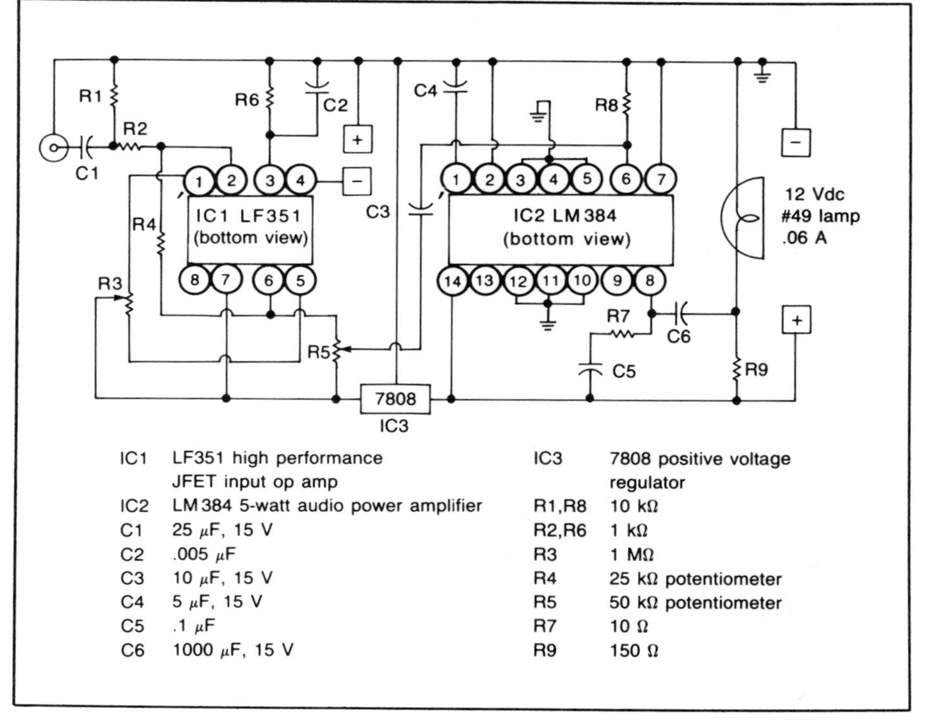

Fig. 8-28. Light transmitter (beam wave).

compensated with R7 and C5 to prevent self oscillation. A #49 lamp that is rated at 2 V and 60 mA can be easily driven. The light beam must be set at a threshold level with R9 to keep the lamp in the on state before modulation. An audio signal generator is utilized to make initial adjustments. The reflector will need to be adjusted to produce a narrow spot of light that is aimed at the receiver. Pins 3,4,5, and 10,11,12, can be grounded for heatsinking.

Further Study

Try a tape recorder input and modulate the beam with music.

PROJECT 65
LIGHT BEAM RECEIVER

Function

The light beam receiver is a companion unit to the light transmitter. It's an example of communicating by light whether in air or through fiberoptics media. (See Fig. 8-29.)

IC1, LF351, is a high-performance JFET input operational amplifier with gain determined by the R9/R6 ratio (or 1 megohm/1 kilohm in this circuit) to provide a gain of 1000. The light signal modulation needs considerable drive, therefore, a transistor preamplifier is in the circuit through Q1. The operational amplifier IC1 must have close to zero reference voltage at pin 6 to ensure maximum gain. Pins 1 and 5 provide offset balance via R10. Set a digital multimeter between pin 6 and B+ and then adjust R10 to read close to zero at no signal input.

The bias of Q2 is adjusted via R3 and R4 to establish quiescent voltage of 4.5 between ground and emitter resistor R1. Phototransistor Q2 is an emitter follower configuration. Light modulation received by Q2 is detected as audio voltage at the base of Q1. The output voltage from Q1 is coupled via C2 to the operational amplifier via volume level control R7.

The output signal from IC1 is coupled to an outboard amplifier

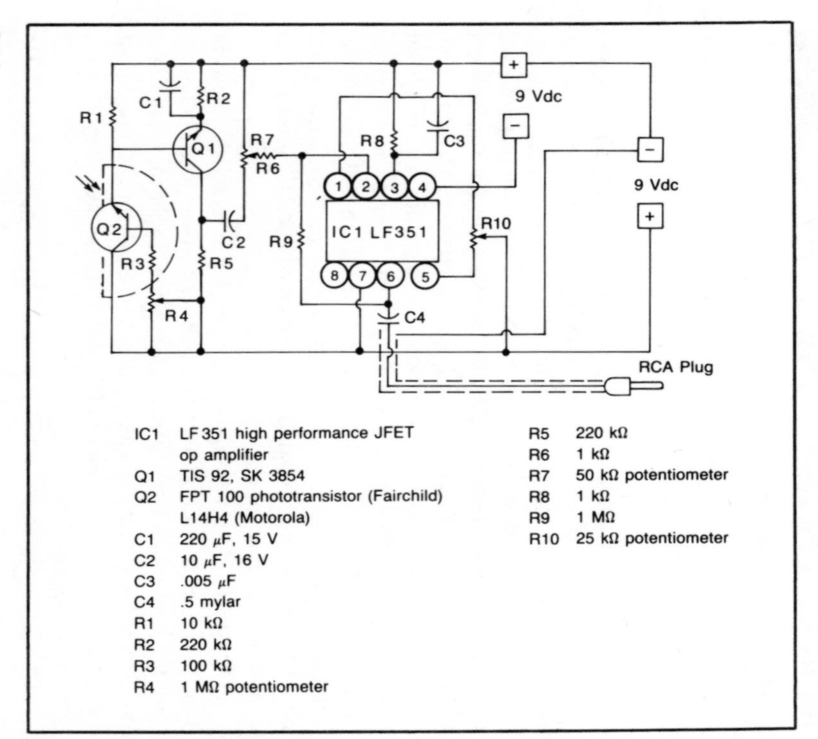

Fig. 8-29. Light beam receiver.

via C4 and the RCA plug. It is important to use good audio shielding techniques to reduce hum level in the final amplifier.

Further Study

You can use a flash light parabola (mirror) without lens as the light catching source for the transmitted light. You will need to adjust the phototransistor inside the parabola to find the focal length. Remember the phototransistor window is reversed in order that it faces the parabolic mirror. You can use a fiberoptics cable between transmitter light and the phototransistor on the receiver as the carrier of the signal.

PROJECT 66
DUAL EMERGENCY FLASHER

Function

A dual emergency flasher (amber light) provides two alternate flashing lights to give the impression of a light moving back and forth from two positions. This system provides greater visibility to warn of a dangerous situation. (See Fig. 8-30.)

IC1, LM384 power audio amplifier, is designed as a power oscillator with a frequency of about 5 hertz per second. Pin 2 has a reference voltage via R3 and output pin 8. When the unit is turned on, output pin 8 is HIGH and C2 begins to charge via R2 until it reaches the reference voltage. Lamp 1 comes on at turn-on also. When the reference voltage is reached, output pin 8 goes LOW and lamp 1 goes off. Since pin 8 is LOW and lamp 2 is connected to the B+ from pin 8, it turns on. It will remain on until C2 discharges to lower the voltage at pin 6 at which time output pin 8 swings HIGH to complete one hertz.

The output is symmetrical or about 100 microseconds per time-on for each lamp. The lamps are 6.3-volt types that are powered by a 12-volt supply, therefore the burst provides a short overload to give a higher intensity flash. The LED, D1, flashes at the same rate as lamp 1. It monitors the circuit operation even if one or both lamps

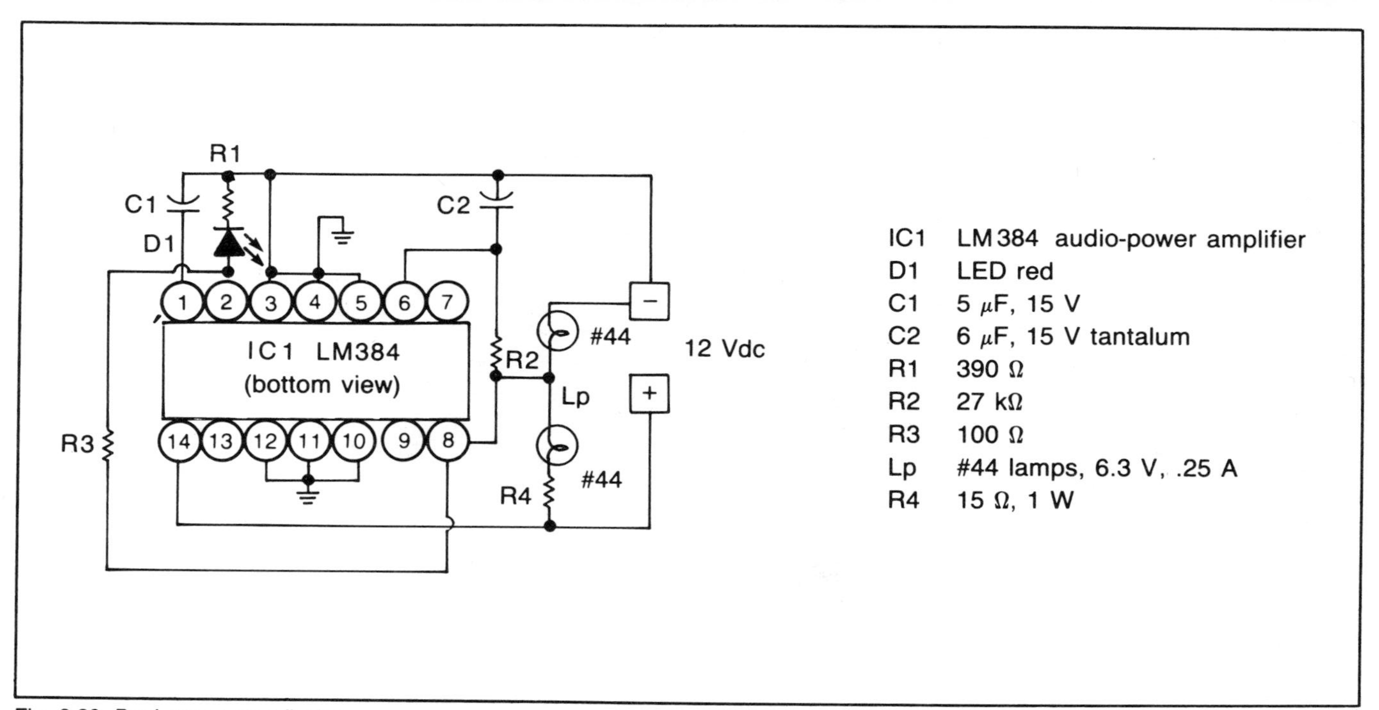

Fig. 8-30. Dual emergency flasher.

should burn out. The circuit will operate on any voltage between 9 to 12, but with reduced intensity at the lower voltages. The decay time of lamp 2 is not as sharp as lamp 1 but the difference is so slight as to be unobjectionable.

The unit can have the outboard lamps mounted in amber pill boxes with a zip cord as extension leads. The main power lead can have extension with zip cord and adaptor plug for quick connection to the automobile cigarette lighter.

The circuit was tested for one hour with no appreciable heat from the IC. It is recommended that pins 3,4,5, and 10,11,12 be connected to a grounded copper foil for heat dissipation.

Further Study

Change the value of R2 and measure the output at pin 8 with an oscilloscope for time on and off; or note difference in alternate flashes. (Do not go above 68 kΩ or unit will stop oscillating.)

PROJECT 67
SAFETY FLASHER (ROAD HAZARD)

Function

A flashing amber light is a caution reminder to others. Changing a tire on a busy highway or an automobile breakdown in a difficult location like over the crest of a hill presents a hazardous situation. Driving in a snow storm with low visibility is another difficult time. A flashing amber light that plugs into the cigarette lighter and is then hung out of the window on the driver side provides an additional safety factor for the motorist. (See Fig. 8-31.)

IC1, LM384, is a power audio amplifier that is designed as a low-pulse multivibrator. A frequency of 5.5 hertz is established with the components C1 and R1. Changing the value of C1 will change the frequency rate. Pin 2 has a referenced voltage via R1. When the circuit is turned on, the output pin 8 is HIGH and C2 begins to charge through R3 to pin 8. When the voltage at pin 6 reaches the reference voltage, the output at pin 8 goes LOW and lamp #44 goes out. It will remain out until C2 discharges to lower the input at pin 6, and then the output swings HIGH to complete one hertz. The *on* time is 60 microseconds, whereas the *off* time is 120 hertz, therefore the pulse is nonsymmetrical.

A high intensity flash is accomplished by utilizing a 6.3 lamp

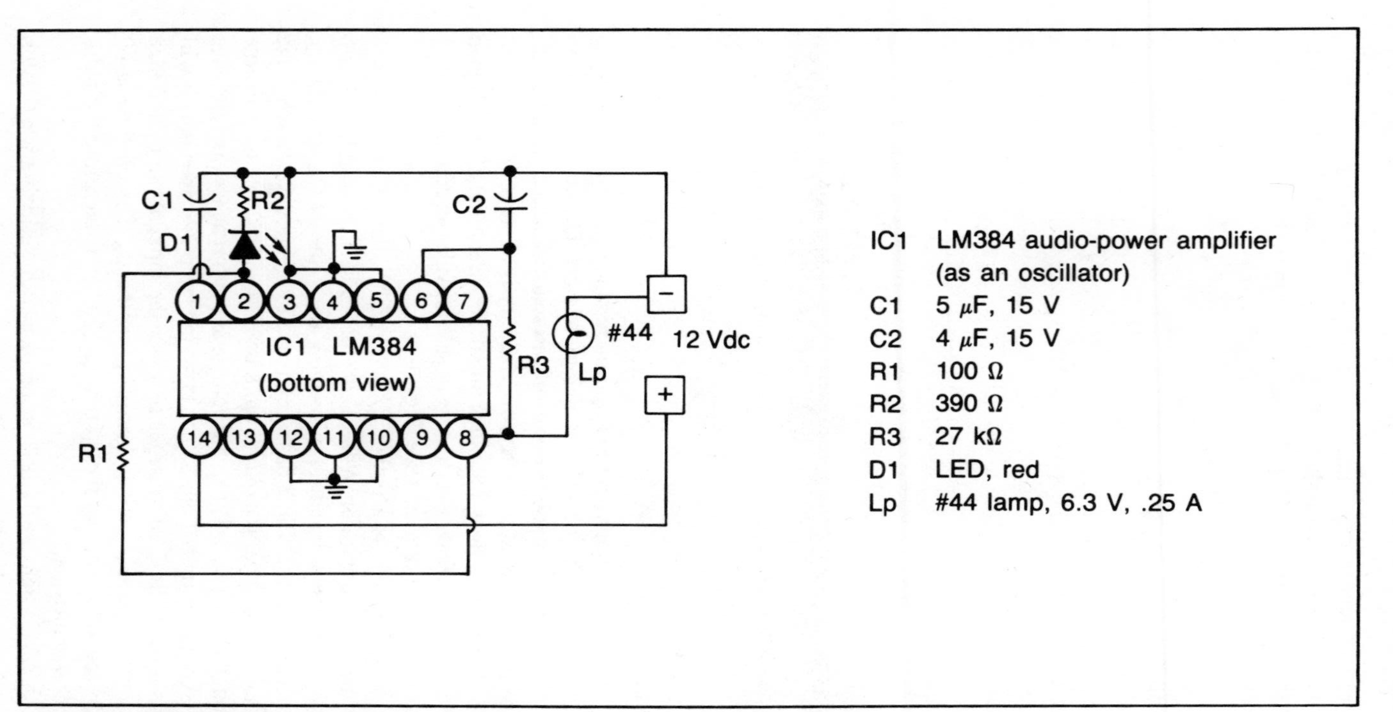

Fig. 8-31. Safety flasher (road hazard).

with a 12-volt supply. The lamp can withstand a short burst of higher voltage for 60 microseconds and not burn out. The LED, D1, flashes at the same rate as the #44 lamp, it monitors the circuit operation even if the #44 lamp should wear out.

The unit can be mounted inside an amber pill box with zip cord as an extension lead. An adaptor plug for the cigarette lighter is available for quick connection.

The circuit was tested for two hours with no appreciable heat from the IC. It is recommended that pins 10,11,12, and 3,4,5 be connected to a grounded copper foil for heat dissipation. The manufacturer states those pins are connected internally to the common for heat transfer to the external sink.

Further Study

Disconnect the lamp and note the continued flashing of the LED. Change the value of C2 slightly (like 1 μF) and note the frequency shift. Measure the output with an oscilloscope and determine the time-on/time-off ratio.

PROJECT 68
CAT AND DOG FINDER

Function

A cat or dog when released to the outdoors at night is sometimes reluctant to respond to your call. The cat or dog finder is designed to help the caller direct his voice toward the animal by identification of position in the dark. (See Fig. 8-32.)

IC1, CD4001 NOR gate, has two gates cross-coupled as a slow speed multivibrator with the oscillator rate determined by C1 and R1. The finish rate is about two flashes per second as seen at LED, D1.

The pulse developed at output pin 3 of the control oscillator is coupled to a third gate, pins 8 and 9, via C2. The truth table of a NOR gate shows that if both inputs go LOW, the output goes HIGH. Output pin 10 of the third gate is connected to the positive supply via LED (D3) and R4; when output pin 10 is HIGH the LED (D3) is out. The end result of this combination is to make LED (D1) and LED (D3) flash alternately.

The output at pin 10 is connected to pins 12 and 13 of the fourth gate via C3. When pin 10 went LOW to turn on LED D3, the state was transmitted to pins 12 and 13 via C3, which made output pin 11 go HIGH to flash LED D2. What one sees is a very slight delay between LEDs D2 and D3. The illusion depending on spacing and

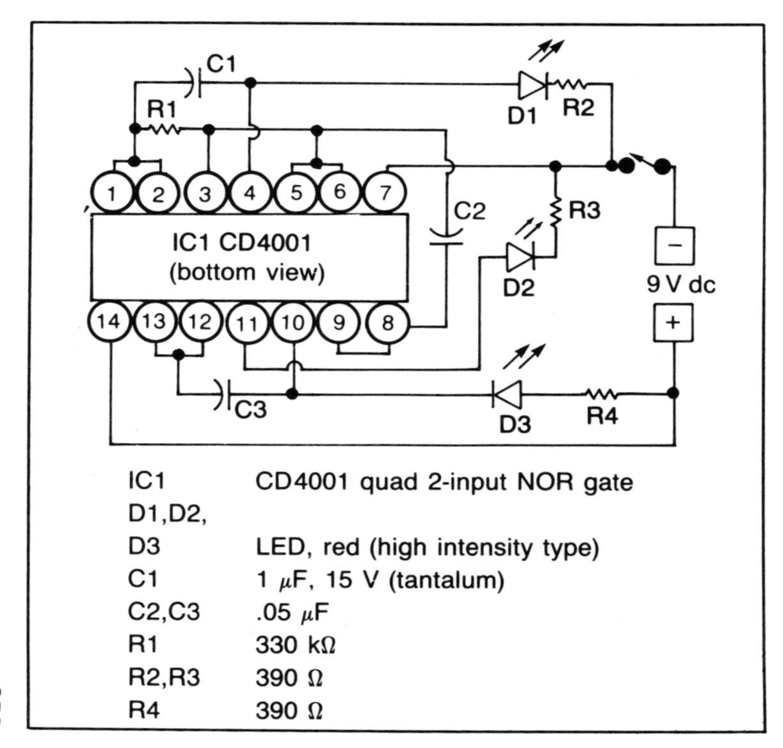

IC1	CD4001 quad 2-input NOR gate
D1,D2,	
D3	LED, red (high intensity type)
C1	1 μF, 15 V (tantalum)
C2,C3	.05 μF
R1	330 kΩ
R2,R3	390 Ω
R4	390 Ω

Fig. 8-32. Cat and dog finder.

location of the three LEDs is that they flash at different times. The LEDs can be spaced on the collar to be seen from right, left, and top position. The visibility distance is a factor of level of darkness. Use high intensity type LEDs.

The device can be packaged in a small space and powered by a 9-volt transistor radio battery. The operational swing current is from 9 to 18 mA with 13 mA as average. The life expectancy of this battery should be good for the short periods the animal is outside. The collar load weight would be minimal.

Further Study

Change the value or C1 or R1 to change the flash rate. The outputs can be connected to the plus or minus supply side with LEDs in corresponding polarity. Try different combinations.

PROJECT 69
THREE-STATE MONITOR/ALARM (FLASHING LIGHT)

Function

The Three-State Monitor/Alarm can be utilized to monitor intrusion at three locations. It can identify the point of intrusion and subsequent entries at other places as the intrusions take place. For example, it can warn one of the entry to a room, the entry to a closet, and finally the entry to a safe. If the inputs are LDRs (light-dependent resistors), it could monitor the presence of light at three locations. (See Fig. 8-33.)

CD4011 is a two-input quad NAND gate. The three gates are utilized to monitor the three positions. The truth table of a NAND gate states that if both inputs are HIGH, the output is LOW. In the standby state, pins 6, 2, and 8 of three gates are held HIGH to the B+. The other pins 5, for input 1, 1 for input 2, and 9, for input 3 are held HIGH via jumper switches to the B+ and also to ground via R1, R2, and R3. The outputs of the gates at pins 4, 3, and 10, are coupled to IC2 enabling pin 4, via R10, R9, and R11. Also, readout LEDs D1, D2, and D3 are in the outputs and will turn on when the outputs go HIGH since they are tied to the B− via R7, R6, and R5.

If the intruder breaks the jumper, which can be a switch or a

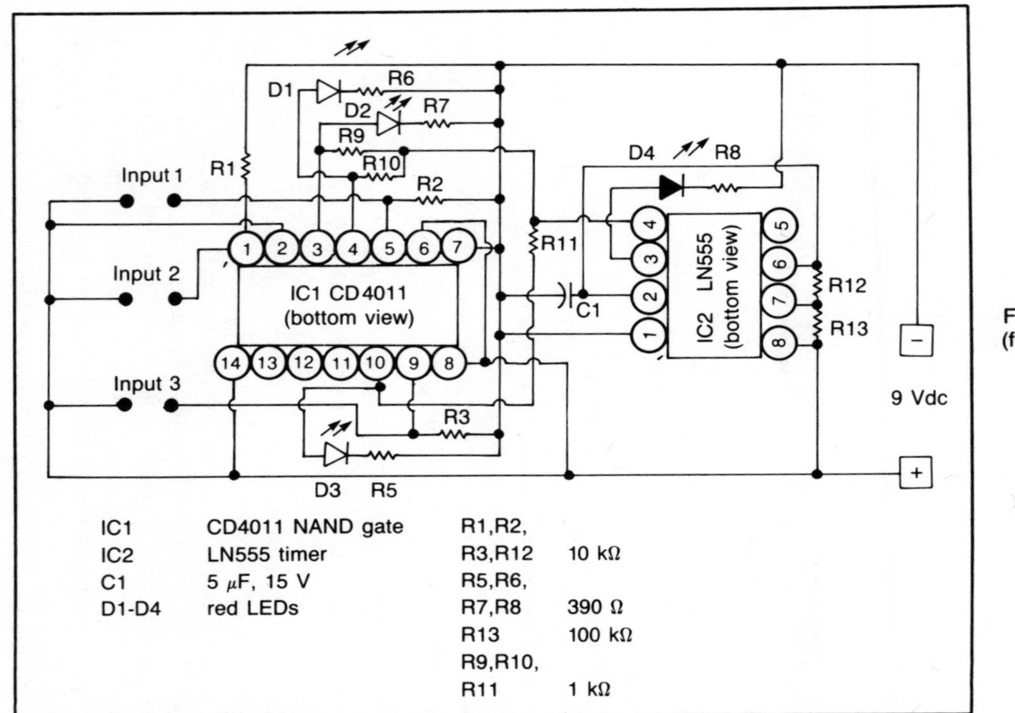

Fig. 8-33. Three-state monitor/alarm (flashing light).

fine wire, the pin goes LOW, thus the truth condition is met to make the output go high. This causes the LED to come on and at the same time turn on the IC2 oscillator via pin 4.

The LN555 timer is a slow pulse (square wave) that alternately switches output pin 3, to flash LED 4 to warn that an intrusion has taken place. The flash rate is determined by C1, R12, and R13. Increasing the value of C1 will slow the flash rate.

Further Study

A thermistor with a positive coefficient can be utilized to monitor temperature at several locations. The ratio of R1, R2, and R3 to the thermistor resistance should be about 10 to 1 at an ambient temperature (or 1 kΩ). It should read about a megohm at a much higher temperature to function properly.

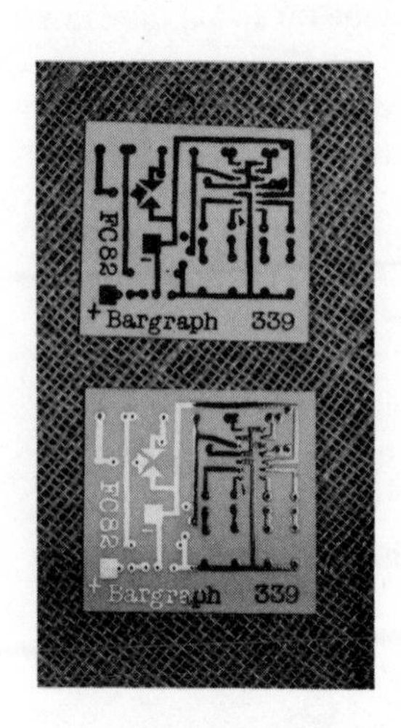

Chapter 9

Power Control Projects

PROJECT 70
ENLARGER TIMER

Function

This enlarger-timer circuit occupies very little space as compared to the usual electromechanical type. It provides time ON of from 1 to 45 seconds, which is ample for an enlarger. (See Fig. 9-1.)

IC1, CD4081, is a quad 2-input AND gate that is the timer. Gate 1 has pin 2 held high, while input pin 2 is held to ground or low through R1. C1 is parallel with R1 and the pin 1 side is connected to S1 that is normally open. When S1 is closed, C1 charges to the supply voltage immediately, thus making pin 1 high and the output pin 3 goes high.

Pin 3 of gate 1 is connected to input pins 5 and 6 of gate 2, which acts as a buffer. When pin 3 is high, output pin 4 of gate 2 goes high. The time that the outputs are held high depends on the discharge time of C1R1. If R1 is adjusted to maximum resistance, the time for C1 to discharge in order to turn off is about 45 seconds. If R1 is set to half resistance value or half way of rotation (linear taper), the time ON would be 22 seconds.

The buffer output of gate 2, pin 4 is connected to the base of a transistor via R2 to bias on Q1. When Q1 comes on it gates the triac which turns on a 100-watt lamp in SO1. R3 is the resistor for

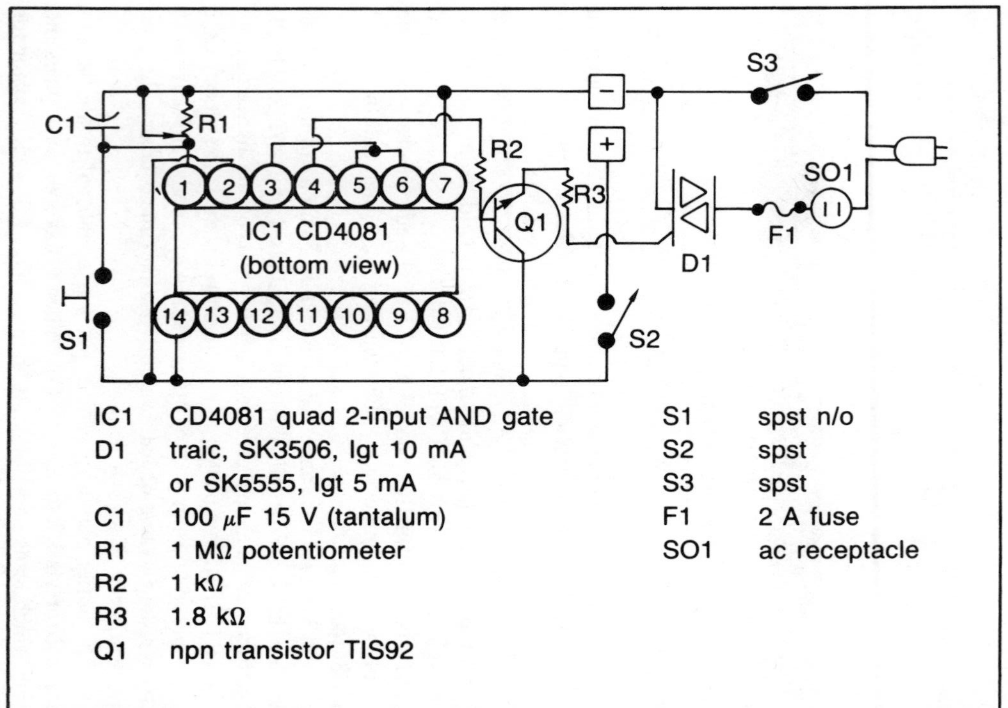

Fig. 9-1. Enlarger timer.

controlling the gate current. It might have to be adjusted up or down in value depending upon the exact characteristics of Q1. When the voltage is removed from the transistor base, the gate current is off and the ac reverse voltage will commutate the triac to turn off the light. The remaining task is to use a pointer type knob for the potentiometer and calibrate the time at different settings.

Further Study

Place an ammeter in series with the 1.8 k gate resistor and measure the gate current. If you continue upward with this resistance you will reach a point where the lamp will be only partly on. Measure the base current of Q1 with SO1 lamp on. You can also reduce the gate current by changing the value of R2.

PROJECT 71
AUTOMATIC NIGHT LIGHT

Function

An automatic night light turns a light on and off by sensing the level of light (ambient). If utilized as a yard light control, it would be on at night and off in the morning. It could be a room lamp that is controlled by outdoor or natural light. (See Fig. 9-2.)

IC1, ULN333OY, is an optoelectronic switch that features several on-chip devices. When light strikes the face side (flat side of a TO 92 case) of the IC, it is in the off state opening the path to ground, pin 2. The output pin 3 is high and D2 is out. Pin 3 is connected to IC2 NAND Gate, pins 1 and 2. The truth condition of a NAND gate states that when both inputs are HIGH, the output is LOW and the transistor is biased off.

When light is shaded from IC1, or darkness comes, pin 3 goes LOW. This causes the LED to light. During the same phase, inputs 1 and 2 of IC2 go LOW, thus making output pin 3 to go HIGH. The transistor Q1 is biased on to provide gate current for the triac and it switches on.

The unit can function on a 9-volt transistor battery for short experimental purposes. It is advisable to utilize a converter for long duration use. Also, IC1 must be shielded from the output light to prevent

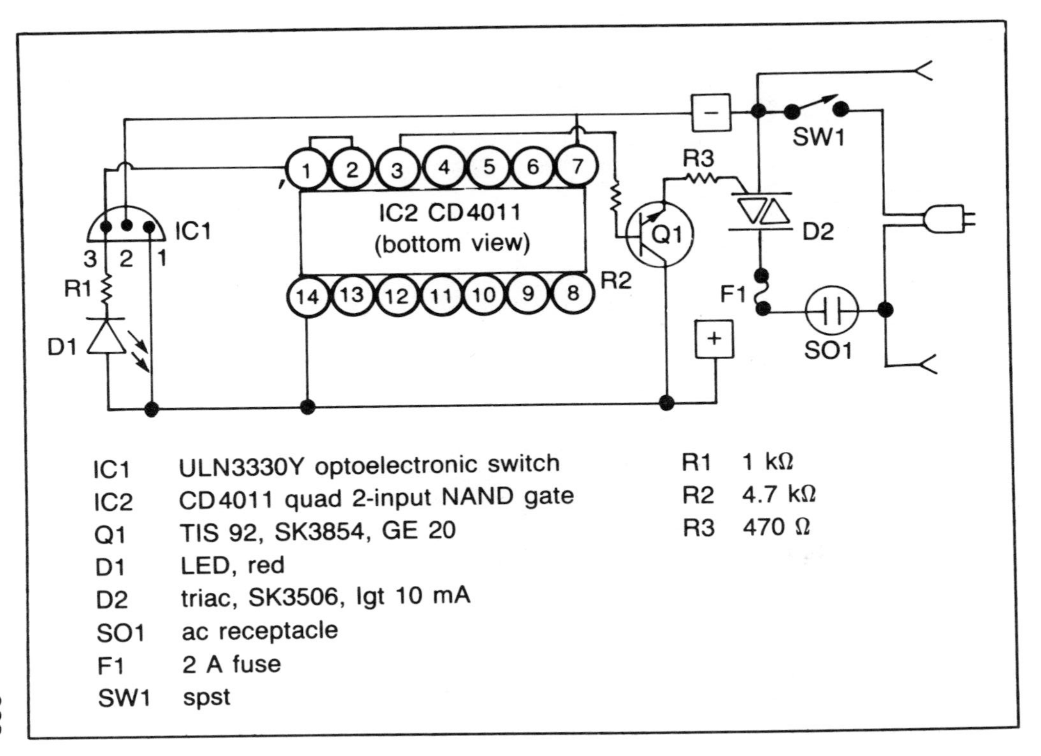

Fig. 9-2. Automatic night light.

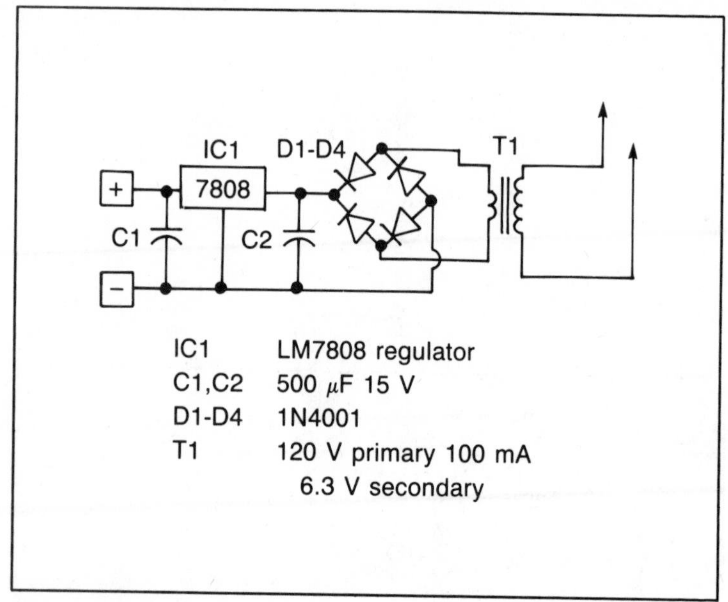

Fig. 9-3. Ac converter for the automatic night light.

light feedback that can have a reverse effect. Figure 9-3 shows a bridge converter for attachment to this circuit.

Further Study

You can build a shroud for IC1 which then requires directional shading to turn on. You can use a pin-hole aperture for light control or even a shutter. Why not build a bridge converter to power the unit?

PROJECT 72
COMPARATOR HEAT DISTRIBUTION CONTROL FAN

Function

Wood burning stoves produce localized heat unless that energy can be distributed to a wider area. The comparator-heat distribution fan control is a circuit that turns on a fan at a certain level of temperature, thus distributing the heat in a wider area for more comfort. (See Fig. 9-4.)

IC1, LM308, is a comparator that turns on Q1 via R7 when output pin 6 goes HIGH. The emitter of the npn transistor is connected to ground through the gate of D3 (triac) which in sequence turns on the ac socket for fan control. Q1 is utilized in the circuit to increase sensitivity; it has a beta of about 100. Q1 shows a gate current through the emitter of 15 mA with a base current of .3 mA. The gate current circuit is emitter, collector, D2 and R8. Since the triac gates on at 10 mA, you could increase the value of R8 from 390Ω to 470Ω.

The temperature control side of IC1 is determined by comparing the input voltage at pin 2 with the reference voltage at pin 3. The reference voltage at pin 3 should be about 4 volts, IC1 output is off and fan is off.

Thermistor R1 must be of the negative temperature coefficient type which means a fall in resistance with an increase in temperature.

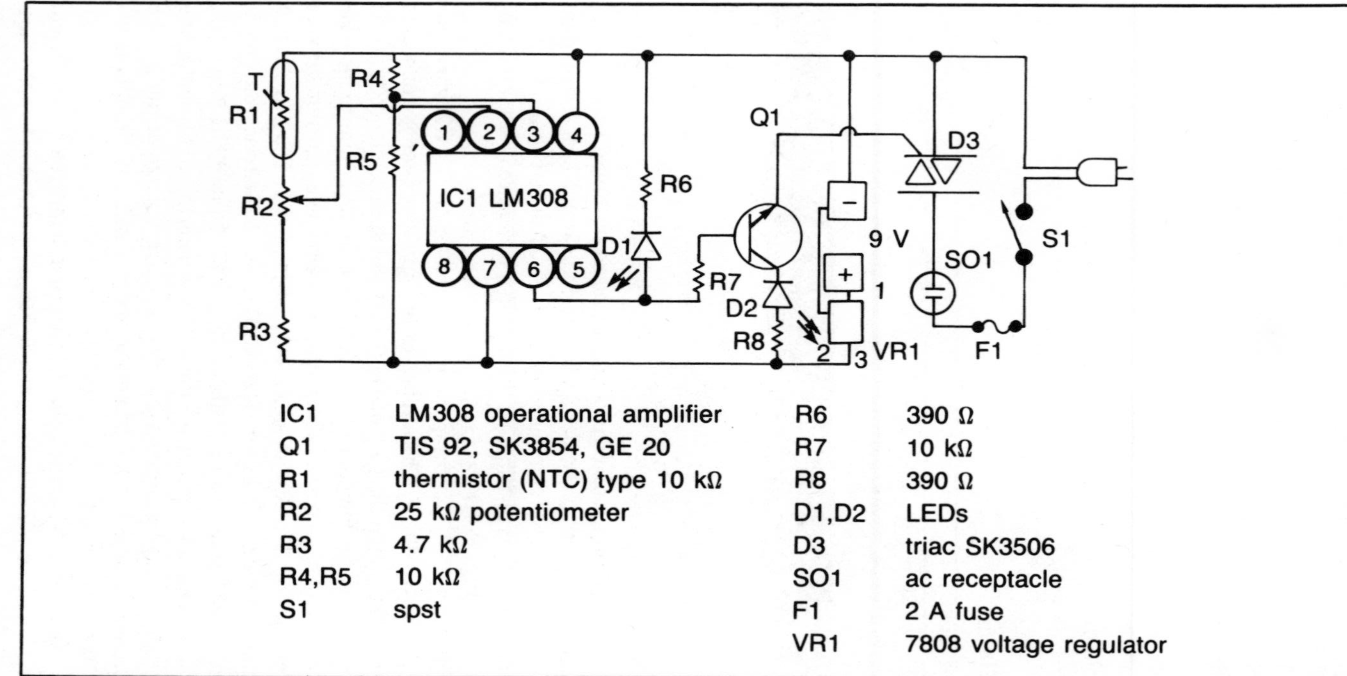

IC1	LM308 operational amplifier	R6	390 Ω
Q1	TIS 92, SK3854, GE 20	R7	10 kΩ
R1	thermistor (NTC) type 10 kΩ	R8	390 Ω
R2	25 kΩ potentiometer	D1,D2	LEDs
R3	4.7 kΩ	D3	triac SK3506
R4,R5	10 kΩ	SO1	ac receptacle
S1	spst	F1	2 A fuse
		VR1	7808 voltage regulator

Fig. 9-4. Comparator—heat distribution control fan.

If a positive temperature coefficient type is utilized in the circuit, one must switch the positions of R1 and R3.

Testing the unit can be accomplished by heating a piece of paper towel against a register and then pressing it against the thermistor. A higher temperature will make R1 drop in resistance, which will cause the voltage drop at pin 2 to go below the reference voltage of 4 volts. One should set the voltage at room temperature at pin 2 to about one-half volt above the reference voltage. VR1 is in the circuit to hold the IC voltages constant for calibration accuracy in the event the 9-volt supply should vary slightly.

Further Study

Change the reference voltage to 1/3 to 2/3 ratio by changing the values of R4 and R5. Try a thermistor with a positive coefficient but don't forget to switch locations for R1 and R3. Remember, you may have to set the unit at several locations in reference to the stove to find the best control pattern of heat circulation.

PROJECT 73
COMPARATOR HEAT FAN CONTROL (DELAY ON-TO-OFF)

Function

A comparator heat fan control have many short OFF and ON cycles when the thermistor is directly in the air stream. This circuit is designed to delay the ON to OFF control cycle even when the comparator calls for shutdown. (See Fig. 9-5.)

IC1 is a comparator in which the output is on and goes off when the voltage at input pin 2 changes from below to above the reference voltage at pin 3. When heat strikes the thermistor (NTC Type), the resistance falls; current increases through R1, R2, and R3 to produce a rising voltage at pin 2. When the rising voltage at pin 2 exceeds the reference voltage at pin 3, the output at pin 6 goes low.

IC2 is a monostable (one shot) multivibrator with a time on delay to off determined by time constant C3, R6. The time with components shown is about 1.25 minutes. The falling voltage at pin 6 of IC1 is coupled via C3 to the trigger pin 2 of IC2. Output pin 3 of IC2 goes HIGH and thus providing bias voltage via R7 to base of Q1.

Transistor Q1 gates on the triac and SO1 becomes energized to turn on the fan. If the fan reduces the temperature at the thermistor and the voltage at pin 2 falls below the reference voltage, the output goes back to HIGH, but it has no effect on the monostable until it has

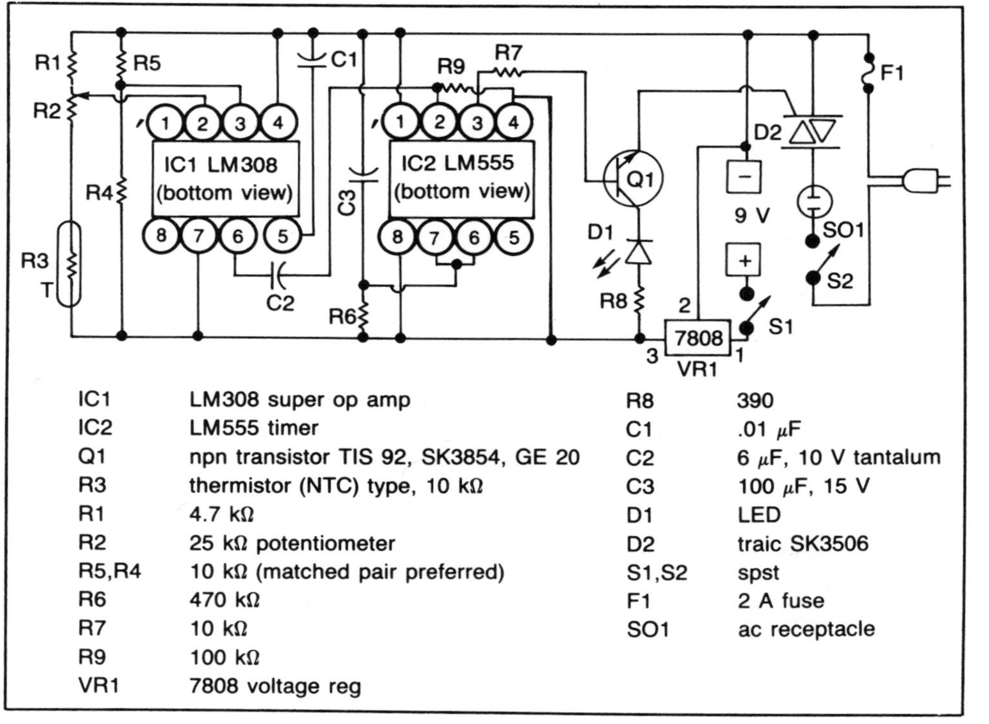

IC1	LM308 super op amp	R8	390
IC2	LM555 timer	C1	.01 μF
Q1	npn transistor TIS 92, SK3854, GE 20	C2	6 μF, 10 V tantalum
R3	thermistor (NTC) type, 10 kΩ	C3	100 μF, 15 V
R1	4.7 kΩ	D1	LED
R2	25 kΩ potentiometer	D2	traic SK3506
R5,R4	10 kΩ (matched pair preferred)	S1,S2	spst
R6	470 kΩ	F1	2 A fuse
R7	10 kΩ	SO1	ac receptacle
R9	100 kΩ		
VR1	7808 voltage reg		

Fig. 9-5. Comparator—heat fan control—delay ON-to-OFF.

run the set time. If on the other hand, the temperature is still high and output pin 6 is low, it becomes the override and the fan continues to run until the thermistor resistance increases at reduced temperature. At this point the one shot will shutdown.

Transistor Q1 is used to increase sensitivity rather than direct couple of gate to pin 3 of IC2; however, one could couple directly from gate of triac to pin 3 of IC2. The LED D1 is in the circuit to monitor the switching action.

Further Study

Change C3 or R6 to a higher value to increase the on time of IC2. Remember you can calibrate with R2 the specific turn on times at various temperatures at the thermistor. You may use thermistors with different resistances and temperature linearity specifications.

PROJECT 74
SUMP PUMP SWITCH (WITH DELAY ON-TO-OFF)

Function

The Sump Pump Switch is designed with a delay of on-to-off in order to prevent short recycle periods. The pump capacity, sump volume, and electrode spacing must be considered in planning the time on-to-off delay. (See Fig. 9-6.)

IC1 is an AND gate in which two gates are utilized in the control circuit. When water enters the sump and rises to the upper electrode, pin 2 goes HIGH and C1 charges to full voltage. Since pin 1 is already HIGH, the condition of both inputs HIGH has been met and output pin 3 goes HIGH. The second gate acts as a buffer and its inputs go HIGH as a result of connection to pin 3. When these inputs go HIGH, output 4 goes HIGH and turns on transistor Q1 via bias resistor R2 and the triac is gated on to turn on the sump-pump motor.

When the water is lowered below the top electrode, pin 2 at the first gate has the supply voltage removed; however, C1 begins to discharge through R1 and thus holding pin 2 to the HIGH side to keep all gates on. It takes a little over a minute to discharge from 9 V to 3.75 V at which time the pump is turned off. Changing the values of either C1 to R1 will change the length of on delay to off. The ground

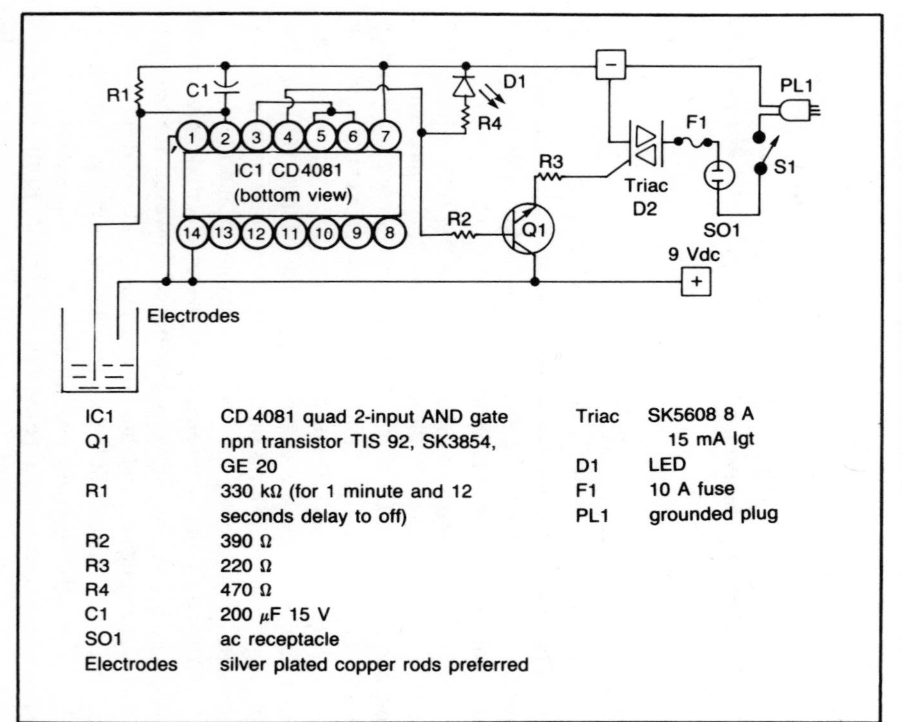

IC1	CD 4081 quad 2-input AND gate	Triac	SK5608 8 A 15 mA Igt
Q1	npn transistor TIS 92, SK3854, GE 20	D1	LED
R1	330 kΩ (for 1 minute and 12 seconds delay to off)	F1	10 A fuse
R2	390 Ω	PL1	grounded plug
R3	220 Ω		
R4	470 Ω		
C1	200 μF 15 V		
SO1	ac receptacle		
Electrodes	silver plated copper rods preferred		

Fig. 9-6. Sump pump switch (with delay ON to OFF).

side of the ac line must be connected to the minus side of the dc power input for safety reasons (be sure to use a grounded plug).

Further Study

Change the value of C1 to 100 μF and note the shorter on-to-off period. Look up the pumping capacity of the Sump Pump and calculate the amount of water to be discharged in one minute delay in on-to-off. Calculate the volume of the sump container and determine how far the water will drop during that time assuming the water inlet was off in that period of time.

PROJECT 75
TIME DELAY ON SWITCH (WITH LATCH)

Function

A deterent to theft could be a time-delay to on-switch that turns on a light or an alarm unexpectedly a short period of time after entering (it latches—thus reclosing the entrance opening has no effect). The device could be utilized for other safety monitoring situations. (See Fig. 9-7.)

One half of the quad AND gate, pins 1 through 6, are utilized as a time delay off-to-on self-latching switch. The truth table of an AND gate states that both inputs must be HIGH to make the output go HIGH. Gate 2 pins 5 and 6 are held high; pin 6 directly to the positive supply and pin 5 via Q1 base to emitter, therefore output pin 4 is HIGH. Since pin 4 is connected directly to pin 2 of gate 1 and pin 1 is held HIGH via S1, output pin 3 is HIGH. In the set condition all pins are HIGH. Note that transistor Q1 is in reverse bias and consequently in cut-off state.

When S1 is opened as a result of someone opening a door, C1 begins to charge via R1, thus continuing pin 1 HIGH but slowly falling in voltage as it comes toward full charge. After seven seconds delay, pin 1 goes LOW to make output pin 3 go LOW; pin 5 connected to pin 3 goes LOW. Output pin 4 goes LOW and pin 2 goes with it

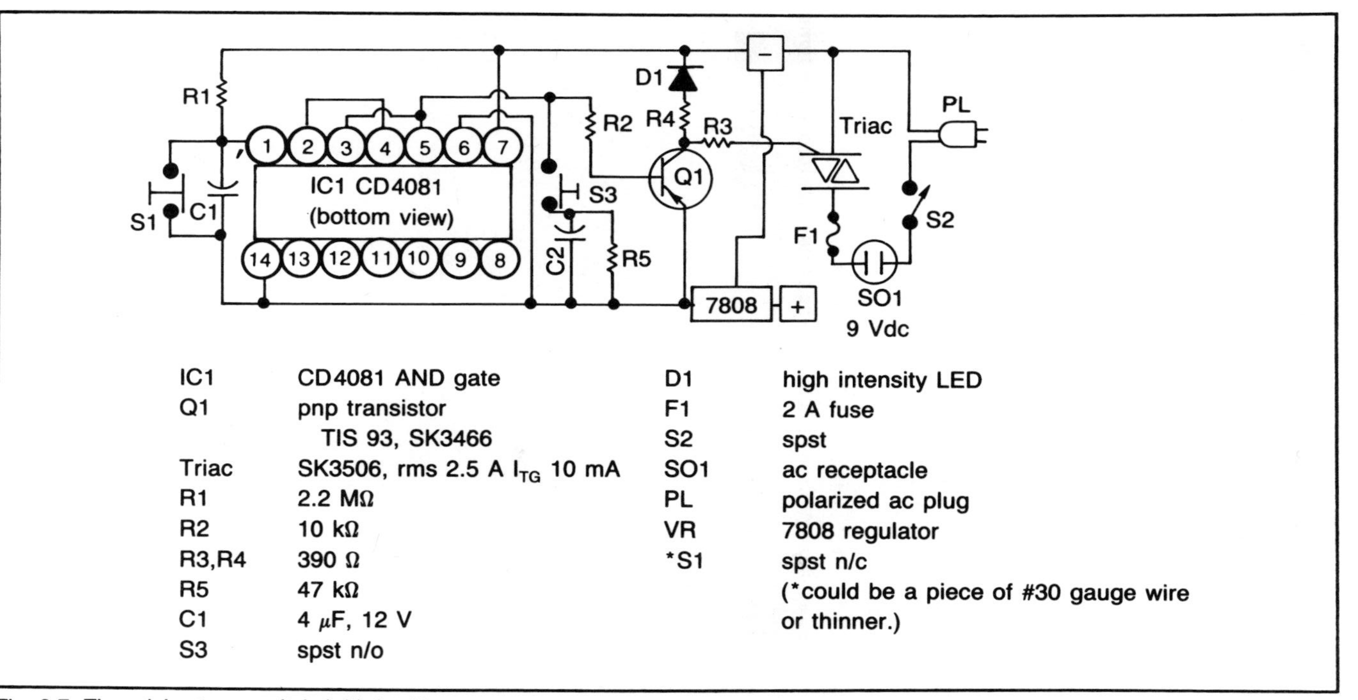

Fig. 9-7. Time delay-to-on switch (with latch).

by direct connection. If an intruder was to close the door thus activating S1 to make pin 1 reverse to HIGH, it would have no effect on the output, since pin 2 is still low, thus we have a latched state.

Back when pin 5 went LOW it biased Q1 on, which gated the triac via R3 and the ac receptacle was energized. The LED in the collector leg will come on also as a monitor. You can use a radio, light (strobe) or some other 120 Vac device in the receptacle.

A reset of the latch is made by pressing S3 to momentarily switch pin 5 HIGH via C2. Remember to reset S1 at the same time. Component R5 is in the circuit to discharge C1 to get ready for the next reset.

Delay time can be lengthened by increasing the value of R1 or C1. Be sure to use a polarized ac plug and connect it in a way to make the ground side to the minus side of the circuit.

Further Study

Change the value of C1 to 10 μF and compare the delay off state to on. Remember it will not latch until the circuit has turned the light on, therefore a good practice is to keep the delay time short.

PROJECT 76
MAGNET ACTIVATED SWITCH (AC POWER)

Function

A solid-state switch that is activated by a magnet is an excellent interface for ac power control. A small ferrite magnet going toward or across the face of the IC will activate the switching action. (See Fig. 9-8.)

The action is based on the Hall-effect concept, which states that when a magnet is brought close to an electron field of a solid-state material, the electrons move away from the magnetic field thus compacting at one side to increase electron mobility. This has the same affect as decreasing resistance and thus increasing current.

IC1 when energized via pins 1 and 2 has an output at pin 3 that is high. Since LED1 is connected to the positive supply and the cathode end to the output pin 3 via R1 that is positive also, the LED is out. When a magnet is brought close to IC1, the output pin goes LOW and the LED turns on. The north pole of the magnet must be to the circle spot reference on the IC face.

One half of IC2 (NOR gate) provides interface to the triac to control the ac line voltage. The truth condition of a NOR gate states that when either input is HIGH the output is LOW or when both inputs are LOW, the output is HIGH. Only one gate would be necessary to ac-

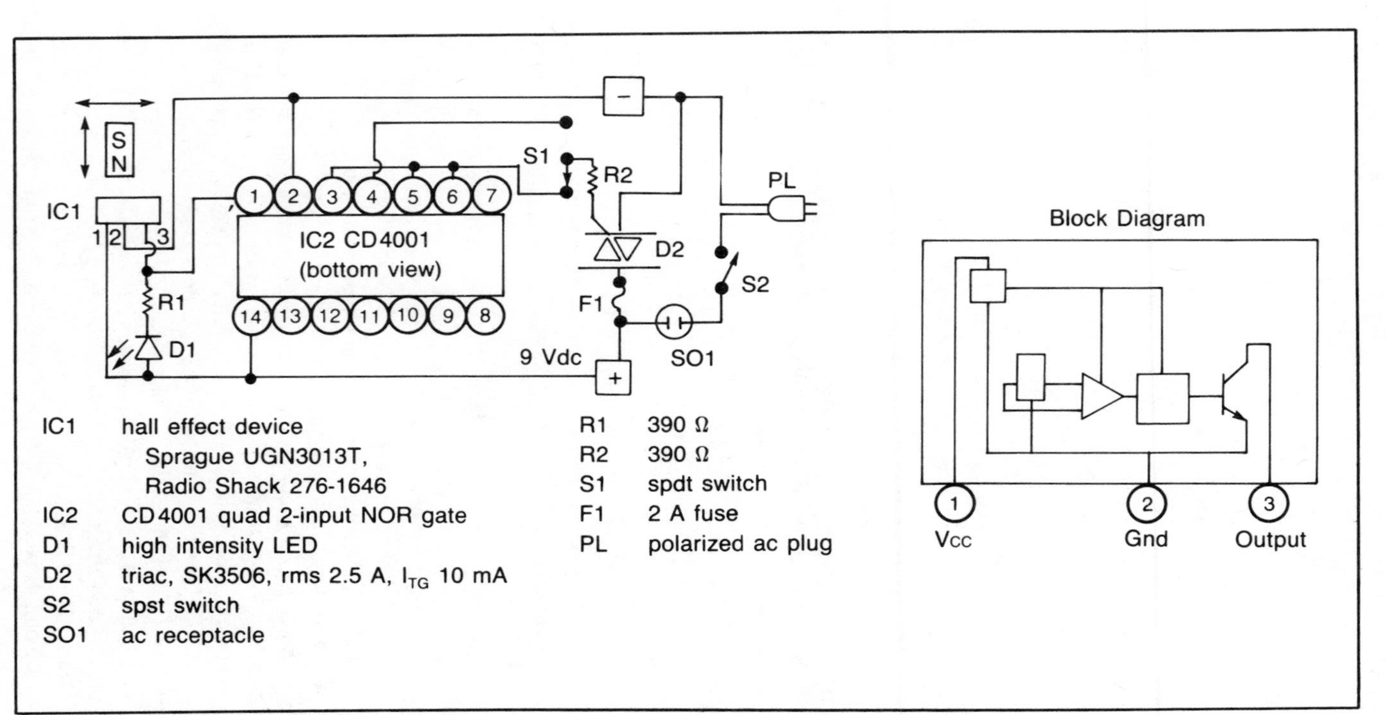

Fig. 9-8. Magnet activated switch—ac power.

tivate the power switch; however, by using an additional gate, you get another option of a reversed switching state.

In the off-to-on operation, IC1, pin 2 is LOW via direct connection to the minus voltage while pin 1 is HIGH, via IC1 output. This mode makes output pin 3 LOW and D2 (with no gate voltage) is in the off state. When a magnet activates IC1, output pin 3 goes LOW and thus the LED comes on. However, pin 1 of IC2 goes LOW and the truth condition makes output pin 3 go HIGH to gate the triac via S1 and R2. The ac receptacle is then energized.

Note that output pin 3 of IC2 is connected to both inputs, pins 5 and 6 of gate 2. Output pin 4 of the second gates goes LOW as a result of pins 5 and 6 going HIGH. If S1 had been switched to the top, the triac that would have been on would switch off (thus a reversed option). What one sees in activating IC1 with a magnet is (1) light comes on when S1 connects the gate of the triac to pin 3 or (2) light that was on will go off if S1 connects the gate of the triac to pin 4.

Be sure to use a polarized plug to connect the neutral side of the ac line to minus. Remember to use caution when handling 120-V ac line power.

Further Study

You can utilize another gate on the other side of IC2 along with R2 and another triac to control two lights.

PROJECT 77
TOUCH SWITCH (AC POWER)

Function

A light or other device can be turned on or off by a light touch. Q1 is a pnp Darlington transistor with a gain of 20,000. When coupled to another transistor, the overall gain is the product of the gain of each unit. It is not unusual to get a two stage gain of 200,000. This may seem enormous and it would be under ordinary circumstances but the ac input via finger touch is just a few microvolts. The purpose of amplification is to produce a positive transition pulse to the flip-flop to activate Q1 that drives a triac to energize the receptacle. (See Fig. 9-9.)

The injected ac is amplified via Q1 and Q2 when the end of R1 is touched. The output voltage across R3 is coupled to the input of the flip/flop, pin 3 via C2. The capacitor C1 is in the circuit to bypass the ac component and thus produce a more positive pulse at pin 3 without a ripple. The set, reset, J, and K pins are tied to their respective minus and positive voltages to enable the flip/flop to function in bistable mode. One touch should produce an on state and the second touch to an off state. The touch should be a rapid tap.

The LED is in the circuit to monitor the change in the flip/flop. Q3 is in the circuit to increase sensitivity and decrease loading on

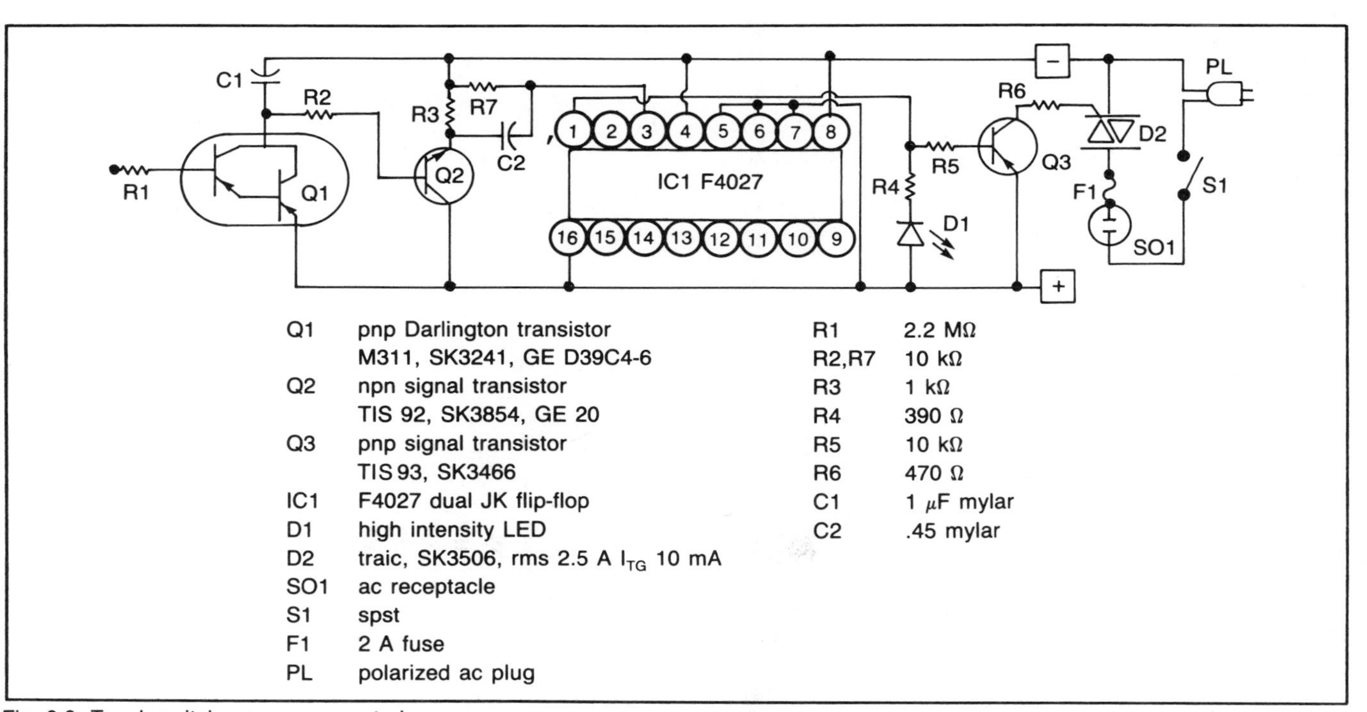

Q1	pnp Darlington transistor
	M311, SK3241, GE D39C4-6
Q2	npn signal transistor
	TIS 92, SK3854, GE 20
Q3	pnp signal transistor
	TIS 93, SK3466
IC1	F4027 dual JK flip-flop
D1	high intensity LED
D2	traic, SK3506, rms 2.5 A I_{TG} 10 mA
SO1	ac receptacle
S1	spst
F1	2 A fuse
PL	polarized ac plug

R1	2.2 MΩ
R2,R7	10 kΩ
R3	1 kΩ
R4	390 Ω
R5	10 kΩ
R6	470 Ω
C1	1 μF mylar
C2	.45 mylar

Fig. 9-9. Touch switch—ac power control.

the flip/flop. The base current is .5 mA and the collector current is 17 mA, which is more than adequate for the triac with a gate current of 10 mA.

Again, use a polarized plug for PL and be sure the neutral side of the line is connected to the minus. Use caution in working with ac line power.

Further Study

Can you replace IC1 with a SCR to produce a touch turn-on and latch circuit? It would be an excellent sensitive theft alarm.

PROJECT 78
AC OPERATED SWITCH (FLUID DETECTOR)

Function

An ac-operated fluid detection switch could be valuable to turn on a small sump pump. This circuit incorporates an optocoupler-switch for line isolation. An override is available by blocking the window of the optocoupler and thus causing a shutdown. (See Fig. 9-10.)

IC1, fluid detector, is operated in dc mode rather than with frequency drive of the output. But adding C1 between pin 12 and minus, the output pin 1 becomes dc rather than ac. However, it is important to keep C1 at 150 μF or larger, otherwise a ripple frequency of the internal oscillator will appear in the output. Remember the internal oscillator is a necessary function for the probes to prevent plating action.

The output pin 1 is high when one of the probes is out of the fluid. The high positive output is connected to the base of Q1 via R1 to reverse bias the transistor and it is in cutoff state. When water rises to touch both probes, the oscillator signal is coupled to the internal detector via C4 and the output voltage at pin 1 goes LOW to turn on Q1. The optocoupler LED is connected from minus voltage to positive supply via R2 and Q1. It will energize when Q1 turns on.

If the window of the coupler is open (not covered) the phototran-

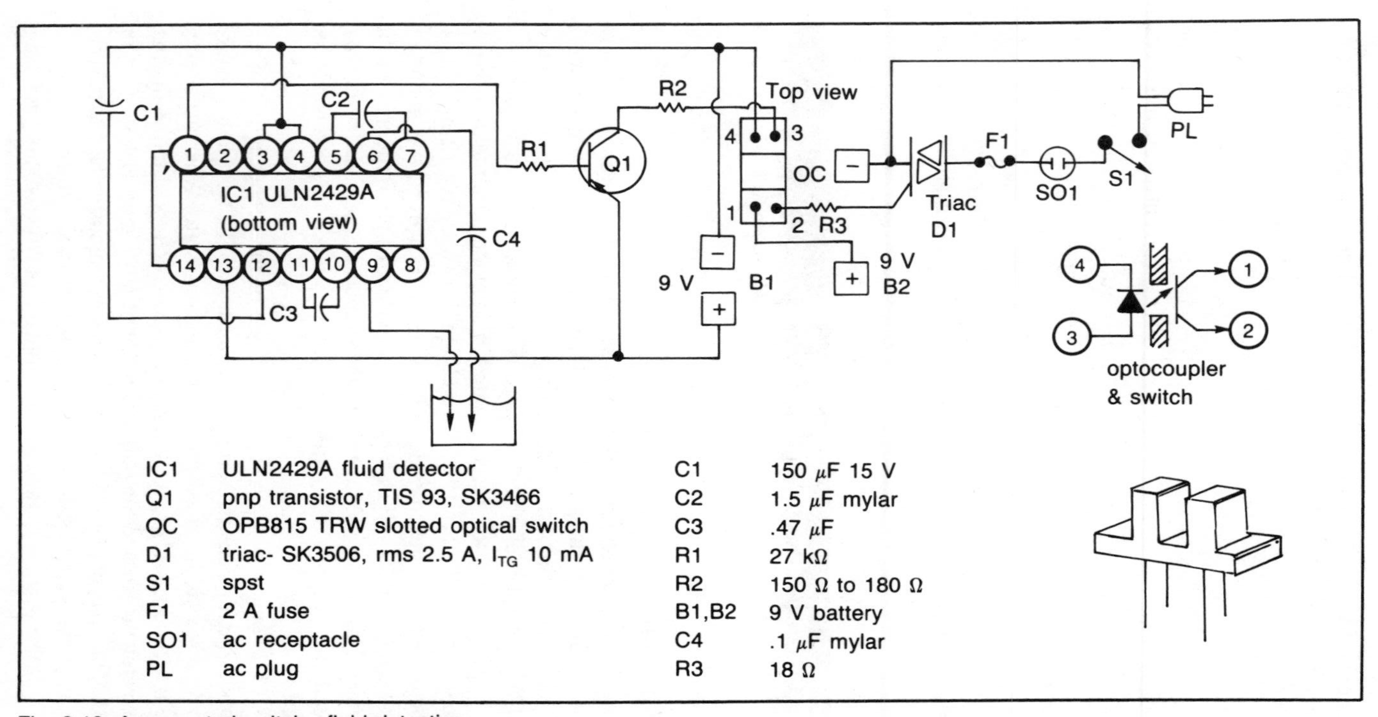

Fig. 9-10. Ac operated switch—fluid detection.

sistor in the coupler will see LED light and turn on. Since the triac gate is coupled to the phototransistor in the optocoupler, it will be energized to activate the ac receptacle. Two battery supplies are utilized to insure complete isolation from the ac line. It is important when working around ground water (if that is the case) to use caution in respect to the HIGH side of the line. The standby current load on battery 1 is 4 mA, whereas the operating current is 50 mA. The operating current on battery 2 is about 11 mA.

Further Study

Activate the circuit by touching both probes to water (use 40-watt lamp in receptacle) and then close the window of the optocoupler to observe override. Remember, the probes can have high/low positions to activate them.

PROJECT 79
OPTICAL SWITCH (AC POWER)

Function

An optical switch can control ac power through an IC interface with a LM311 high performance comparator. A card inserted in the window turns off a light, or a motor driven rotary disk as a shutter can produce a timed sequence of ac control. (See Fig. 9-11.)

The comparator, IC1, has a reference voltage that is set at pin 3 by voltage dividing network R2, R3. Since both of the resistors are equal, the set voltage is 4.5 volts or one-half of the supply. The variable voltage at pin 3 is determined by the optical switch (phototransistor side) pins 1 and 2 plus R3. A closed window of the optical switch keeps light from the phototransistor and thus it is cut off. With no current flowing through the phototransistor and R3, the voltage at pin 3 via the centertap of the potentiometer is nearly zero. The LED side at pins 3 an 4 of the optical switch is on constantly via R7 and the supply voltage. The variable voltage at pin 3 is below the reference voltage and it causes the output at pin 7 of IC1 to be HIGH to keep LED, D1, off. Also, transistor Q1 is reversed bias via R4 and it is cut off. The triac is cut off because there can be no gate current with Q1 off.

When the window of the optical switch is opened, the light from

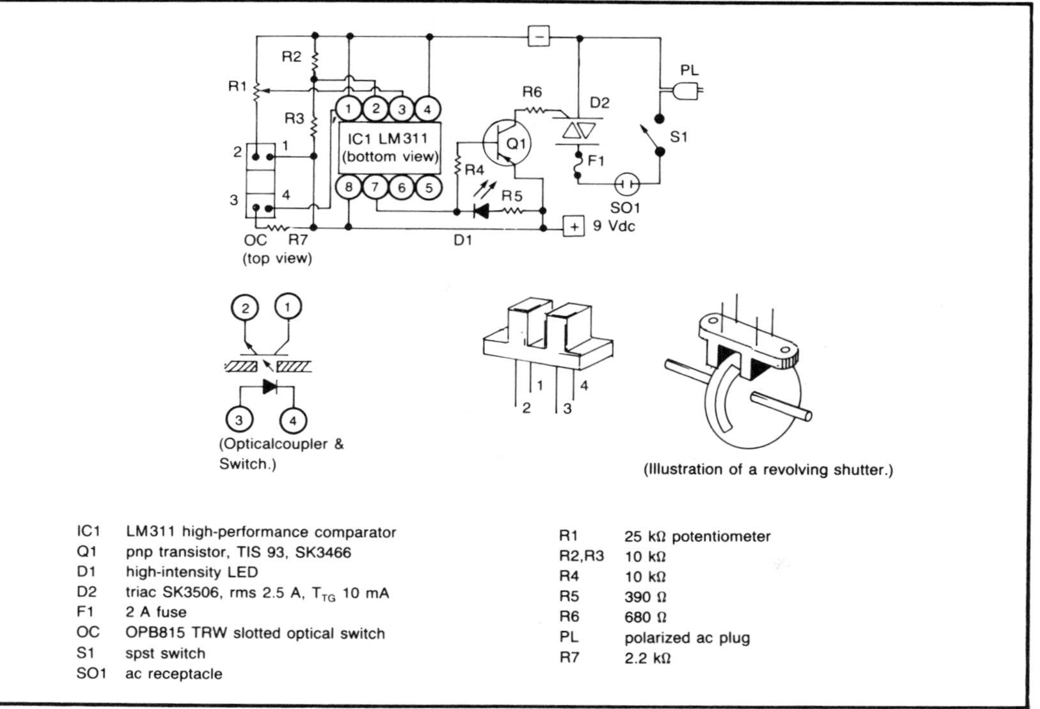

IC1	LM311 high-performance comparator
Q1	pnp transistor, TIS 93, SK3466
D1	high-intensity LED
D2	triac SK3506, rms 2.5 A, T_{TG} 10 mA
F1	2 A fuse
OC	OPB815 TRW slotted optical switch
S1	spst switch
SO1	ac receptacle

R1	25 kΩ potentiometer
R2,R3	10 kΩ
R4	10 kΩ
R5	390 Ω
R6	680 Ω
PL	polarized ac plug
R7	2.2 kΩ

Fig. 9-11. Optical switch—ac power.

the LED side strikes the phototransistor side and it turns on. A current path through the transistor side of the optical switch and R3 causes the voltage at pin 3 to rise above the reference voltage and the output at pin 7 goes low. As a result of pin 7 going low, the LED, D1, comes on. Transistor Q1 switches on, and the triac gates on to energize the ac receptacle. However, R3 potentiometer is an override and you need to adjust it to set the voltage at pin 3 to be above the 4.5 volts of reference when the window is open. If you adjust R3 to produce a voltage below the reference, the ac receptacle will be off even though the window is open.

Standby current from the voltage supply is 5.3 mA while the operational current is 39 mA. This is mentioned to suggest that you would need an ac converted dc supply to power the circuit for prolonged operation, otherwise a 9-volt transistor battery would have a short life.

The window can be closed and opened in many ways such as shown in Fig. 9-11. A clock motor could drive a disk in the window to produce different on cycles. Again, you are reminded to use caution when working with ac line current.

Further Study

Try designing different shutter arrangements to turn the power on and off.

PROJECT 80
TIME DELAY OFF-TO-ON SWITCH

Function

Some amplifiers will produce a "popping sound" when turned on. This is caused by the power supply coming to the quiescent state. The pop effect can be avoided by a delay circuit that turns on power to the amplifier stage some time after the power supply has been on. (See Fig. 9-12.)

IC1, LM3905, is a precision timer that incorporates a reference voltage, a comparator, a latch buffer, logic circuit, transistor reset, and transistor output. When voltage is applied to pin 5 via S1, C1 charges through R1 toward the reference voltage of 3.15 volts. The internal reference voltage (3.15 V) is necessary to insure accurate timing even if the supply voltage should vary. When the capacitor C1 charges to 2 volts, the comparator turns on the output transistor and the collector output at pin 6 conducts to turn on the outboard components, LED monitor, and the relay RL1.

The internal latching circuit holds the output in the conduction state; however, the internal transistor reset comes on to discharge C1 to zero via pin 3 and 4. The circuit remains in conduction until B+, pin 5 is opened via S1.

Changing the values of R1, C1 will change the charge cycle and

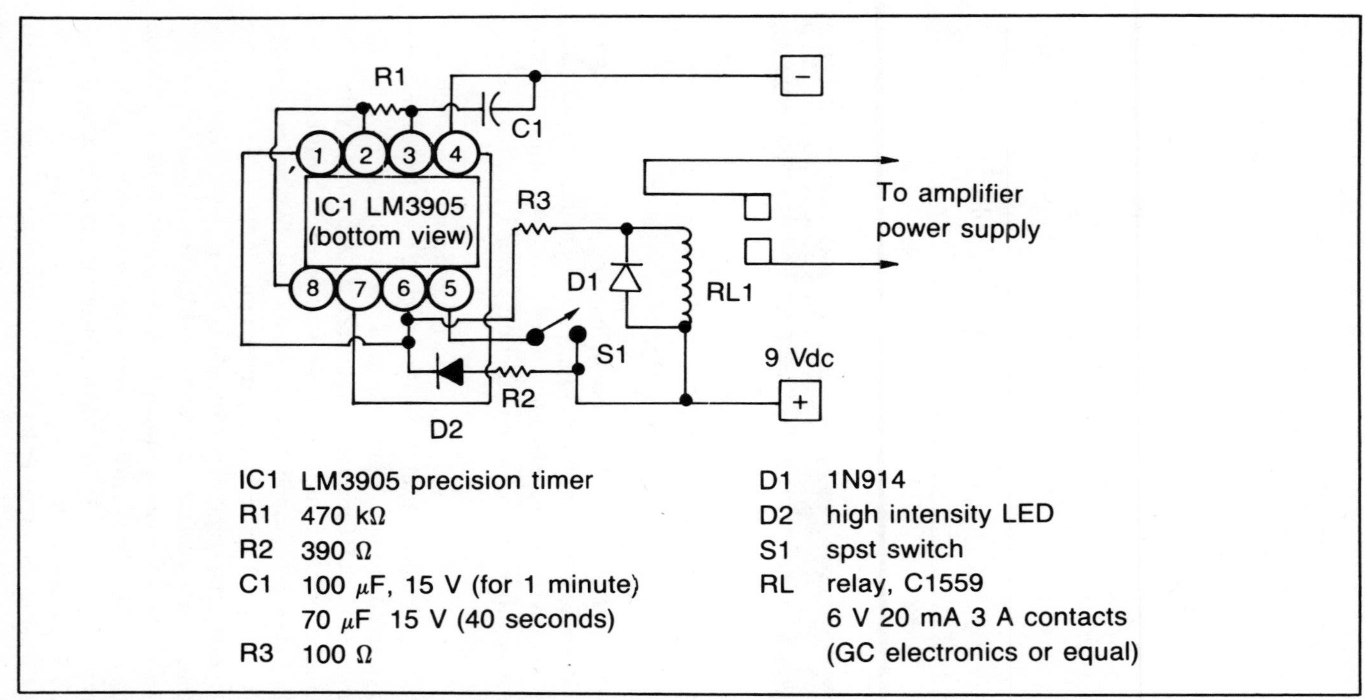

Fig. 9-12. Time delay OFF-to-ON speaker pop reduction system).

the length of time for delay to ON. One can note that 2 volts charge is just about 63 percent of the applied 3.15 reference voltage; thus the time period is one time constant. The exact time can be predicted by using the formula t=RC (.5 × 70 = 35 seconds).

Further Study

Change the value of C1 to 500 μF 15 volts and check the delay with a stop watch. Calculate this time from the above formula.

Ground logic pin 8 instead of connecting it to pin 2 and you have completed a reversal; a delay ON to OFF. When the switch S1 is closed the relay will energize immediately and then go off some time later when C1 charges to 2 volts. It's great for an auxiliary garage light.

PROJECT 81
AC POWER SWITCH (TIME DELAY OFF-TO-ON)

Function

The ac power switch (time delay off-to-on) or (on-to-off) would be useful in several switching applications. A case in point would be to switch a fan on to dry some object and then turn off after a short period. Another possibility would be to switch on an outdoor light just before departing with the car (one that turns off after you have departed). The on-to-off might be a heat lamp with a timed period of on. (See Fig. 9-13.)

IC1, LM3905, is a precision timer with set reference voltage from an internal source. The voltage is 3.15 at pin 2. This is important for accuracy of the timer in the event of supply voltage change. When S1 is closed, C1 begins to charge via R1 and the reference voltage at pin 2. When C1 charges to 2 volts, the internal-transistor collector at pin 6 conducts via R2 and R3 to bias ON transistor Q1. Q1 gates the triac via R4 and the triac energizes socket SO1. The triac gate current of 14 mA is more than adequate because its gate current requirement is 10 mA.

The internal latch of IC1 holds the output transistor in the conduction state and resets C1 to zero. The process is ready to recycle with a shut-off and reswitch of S1.

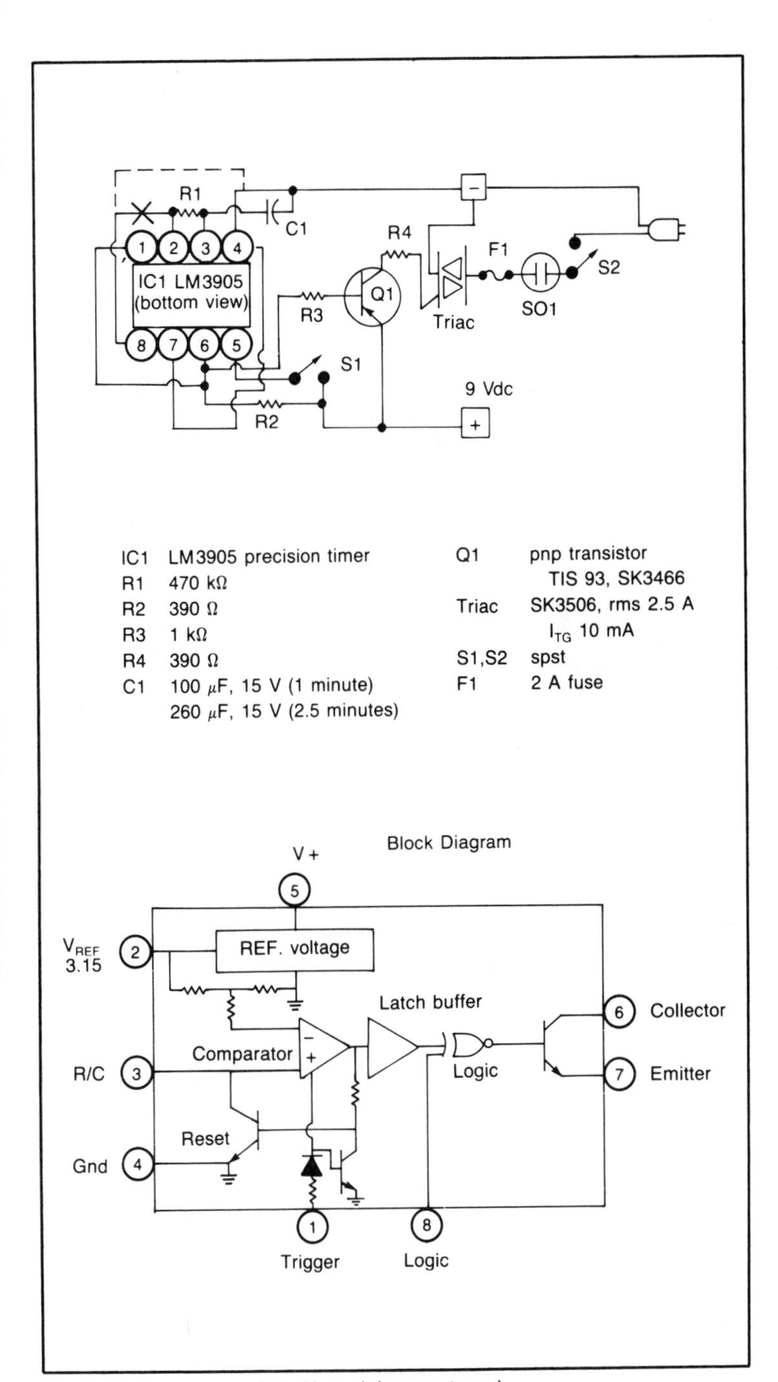

Fig. 9-13. Ac power switch (time delay OFF to ON).

The circuit can operate in reverse mode by changing logic connection pin 8; from pin 2 to ground as shown by the dashed line and *X* separation. In this mode, the light in socket SO1 will come on when S1 is closed and will remain on until C1 charges to 2 volts.

Further Study

Try the reverse mode of operation. Also, change the value of C1 or R1 or both to change the time. Remember (t=RC) for determining the length of on or off time. (R is megohms and C is in microfarads.)

PROJECT 82
AC POWER SWITCH (WITH LATCH)

Function

The self-latching power switch makes an excellent circuit for burglar alarm or other detectors of intrusion. The latch feature of this IC prevents shutdown once the circuit has been activated. (See Fig. 9-14.)

IC1, LM3905, is a precision timer that becomes a "turn on" when a positive transition pulse is applied to the trigger input pin 1 via R1, S1, R2. The internal latching buffer turns on the internal logic circuit, which in turn biases the internal output transistor to on. Since the internal output transistor is connected to Q1, this pnp transistor is biased ON to provide gate current of 14 mA to turn on the triac which energizes the socket SO1. However, the reset switch S2 must ground R/C input pin 3 to keep it below 2 volts or below the 3.2 reference voltage at pin 2 in order to make the latch hold. Once the trigger pulse occurs it relinquishes any further control of the circuit because of the latch.

Opening the reset switch S2 removes pin 3 (R/C) from ground and it rises to the reference voltage of 3.2. Since pin 3 is above the 2 volts, it causes the comparator to reverse the latch and the internal output transistor goes into cut-off. This means that pin 6 of the inter-

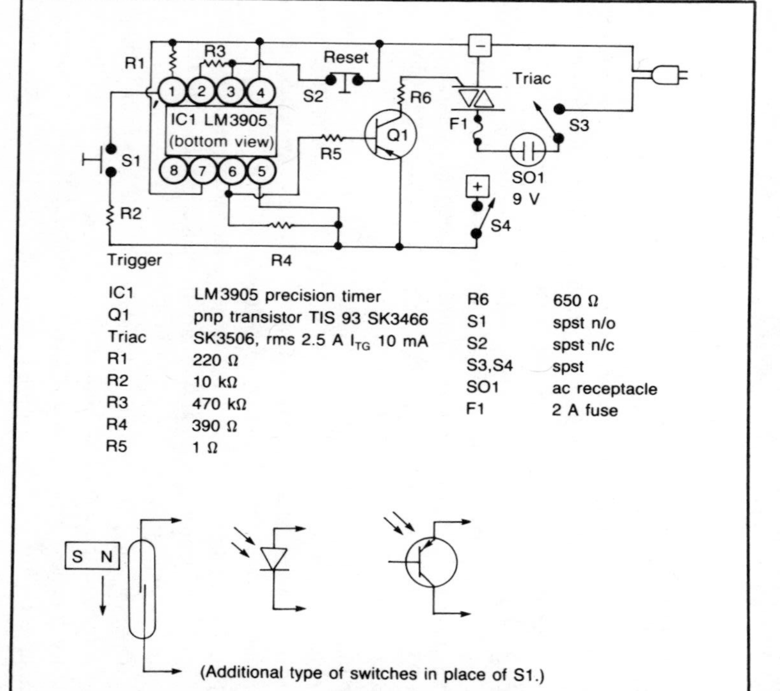

Fig. 9-14. Ac power switch (with latch).

nal output transistor will be at supply voltage to reverse bias Q1 to cut-off and ungates the triac to deenergize socket SO1.

Other switching transducers can be utilized in place of S1 to activate the circuit. Reminder: use caution when working with line ac.

Further Study

Try other transducers to activate the circuit as shown in Fig. 9-14.

PROJECT 83
AC POWER SWITCH
(OFF-TO-ON DELAY AND ON-TO-OFF DELAY)

Function

The ac Power Switch (off-to-on delay and on-to-off delay) would have many applications that depend upon the imagination of the user. One such use would be to turn on a radio or TV at a preset time before the program is to appear and to turn it off after the program is over. The circuit is programmable via changing the values of C1, R1 for off-to-on period and C2, R3 for on-to-off timer. (See Fig. 9-15.)

IC1, LM3905 precision timer, is a retriggerable monostable in off mode with output pin 6 high. When S1 is closed momentarily, the trigger input at pin 1 starts the time cycle. The component C1 charges through R2 toward the reference voltage of 3.1 at pin 2. The output pin 6 goes LOW upon triggering and the monitor LED, D1, comes on. When the charge across C1 reaches 2 volts or 2/3 of the reference voltage, the output goes HIGH again and the LED, D1 goes off. (This is one-shot cycle.)

The output pin 6 of IC1 is connected to the trigger input of IC2, another LM3905 in monostable mode. When the output of IC1, pin 6 goes HIGH at the end of the first time cycle, the positive transition of the pulse will trigger IC2. The component C2 charges through R3

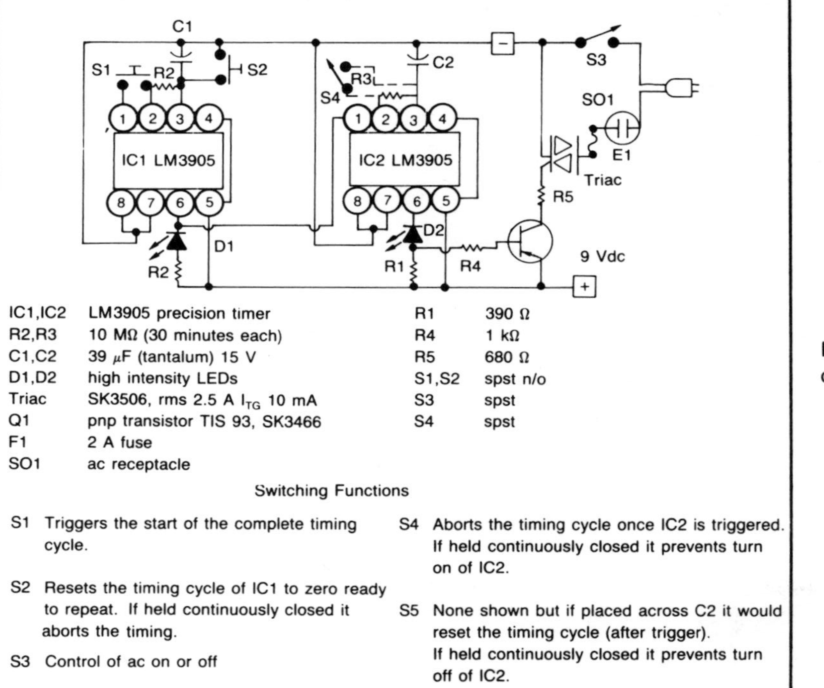

IC1,IC2	LM3905 precision timer	R1	390 Ω
R2,R3	10 MΩ (30 minutes each)	R4	1 kΩ
C1,C2	39 μF (tantalum) 15 V	R5	680 Ω
D1,D2	high intensity LEDs	S1,S2	spst n/o
Triac	SK3506, rms 2.5 A I_{TG} 10 mA	S3	spst
Q1	pnp transistor TIS 93, SK3466	S4	spst
F1	2 A fuse		
SO1	ac receptacle		

Switching Functions

S1 Triggers the start of the complete timing cycle.

S2 Resets the timing cycle of IC1 to zero ready to repeat. If held continuously closed it aborts the timing.

S3 Control of ac on or off

S4 Aborts the timing cycle once IC2 is triggered. If held continuously closed it prevents turn on of IC2.

S5 None shown but if placed across C2 it would reset the timing cycle (after trigger). If held continuously closed it prevents turn off of IC2.

Fig. 9-15. Ac power switch (OFF-to-ON delay and ON-to-OFF delay).

toward the reference voltage of 3.1 volts at pin 2. Upon triggering, the output at pin 6 of IC2 goes LOW and the LED, D2, comes on. The output pin 6 is connected to Q1 via bias resistor R4. Since Q1 is a pnp transistor, it turned on when pin 6 went LOW because it is forward biased. Transistor Q1 gates on the triac to energize socket SO1. When the time cycle is completed in the monostable (one shot), the output pin 6 returns to HIGH again to reverse bias Q1 and it goes into cut/off to turn off the triac.

The circuit can be programmed to change "off-to-on" time and/or "on-to-off" time. Standby current is 4.1 mA for the total circuit. Operational current for IC1 is 23 mA. Operational current for IC1 and IC2 is about 46 mA. The total current is 65 mA for IC1, IC2, Q1, and the triac. The triac gate current is 14 mA. Because of this heavy current draw, it would be wise to construct an ac-to-dc converter supply with a regulator to dc-power the circuit.

Further Study

Logic pin 8 of both ICs is an important control. Try by touching pin 8 of both ICs to their respective pin 2 (and trigger). Try pin 8 of IC1 to its respective pin 2 (and trigger). Try pin 8 of IC2 to its respective trigger pin 2 (and trigger).

PROJECT 84
TIMER-MODEL ROCKET LAUNCHER

Function

A sequential timer provides for adjustable time intervals with LEDs as display of the sequence. It is useful as a model rocket launcher or adaptable for other countdown events. (See Fig. 9-16.)

IC1, LM339, is a quad comparator to turn on four LEDs in sequence. The order of time is at the output pins 13, 14, 1, and 2 in that order. The reference voltage for the comparators is determined by R8, R9, R10, R11, and potentiometer R12. The potentiometer sets the reference voltages and thus the length of time intervals. The input voltages to be compared are applied to pins 10, 8, 6, and 4 by the voltage that is developed across C1 via R13 and S1. When S1 is held in the closed position, C1 charges slowly by the time constant (t=RC) C1R13, which delivers a rising voltage to the inputs. This rising voltage will match each reference voltage in turn and the corresponding LEDs will come on. For example, the reference voltages for one test setting was .33, .66, .99, and 1.32 volts; rising voltage intervals was approximately 3 seconds, or 12 seconds to complete the cycle. One should have the total time cycle for the 4 LEDs to light within 30 seconds for better linearity.

If the reference voltage is set higher by R12, the rising time in-

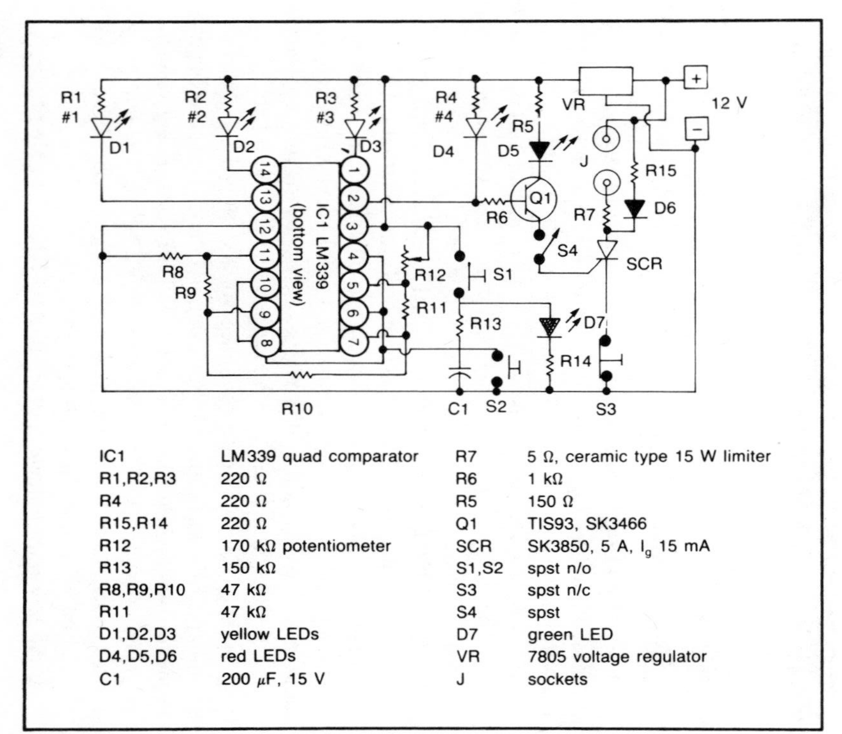

IC1	LM339 quad comparator	R7	5 Ω, ceramic type 15 W limiter
R1,R2,R3	220 Ω	R6	1 kΩ
R4	220 Ω	R5	150 Ω
R15,R14	220 Ω	Q1	TIS93, SK3466
R12	170 kΩ potentiometer	SCR	SK3850, 5 A, I_g 15 mA
R13	150 kΩ	S1,S2	spst n/o
R8,R9,R10	47 kΩ	S3	spst n/c
R11	47 kΩ	S4	spst
D1,D2,D3	yellow LEDs	D7	green LED
D4,D5,D6	red LEDs	VR	7805 voltage regulator
C1	200 μF, 15 V	J	sockets

Fig. 9-16. Timer—model rocket launcher.

tervals will be longer to match each comparator reference voltage. A green LED comes on when S1 is held closed as a monitor of the *go* condition. If S1 is opened at any point in the cycle, the count is in the hold state. Switch S2 is reset to discharge the capacitor C1 in the event of a need to restart.

The 4th LED will come on when pin 2 goes LOW. This will forward bias Q1 transistor that will switch on and thus gate the SCR via S4. Switch S4 is not necessary, however, it does provide another abort option. LED D5 (red) is in the circuit to monitor the gate current turn on.

The SCR applies energy via R7 to the power output jacks that go to the rocket igniter via a 25 foot zip-cord lead. The resistor R7 is a limiter that is in the circuit to protect the SCR in the event the igniter clips at the rocket should short. A red LED, D6 is in parallel with the jack to monitor the energy supplied by the SCR. Note the energy to the rocket is from the 12-V source, whereas the rest of the circuit is 5-volts regulated. VR is in the circuit to hold the voltage constant to insure accuracy of the timer. Remember, a SCR latches on dc, therefore S3 is necessary for power reset to commute the SCR.

Further Study

Use a stop watch and measure several timer intervals as set by R12. Can you add a LM555 as a sound oscillator to provide an audible warning at LED 3 turn-on that warns that the next interval is lift-off?

PROJECT 85
SHIMMERING LIGHT CONTROL (DISPLAY EFFECT)

Function

Variation is light intensity and frequency is attention getting for display purposes. A shimmering light control can be utilized with incandescent candelabrum or Christmas lights to give a flickering effect similar to a natural candle. (See Fig. 9-17.)

IC1, LM555, is a low-pulse oscillator with an adjustable frequency from 10 hertz to 80 hertz via R3. Changing the resistance of R3 will change the length of time of the positive portion of the square wave. By adjusting R3, the oscillator can be made to phase with the 60 hertz ac. Since the SCR operates only on the positive excursion of the sine wave, the light will be approximately at half brightness. In addition, the oscillator positive square wave will gate on the SCR via R4 at some phase of the 60 hertz wave.

The effect is to produce a pulse of energy at the light that will result in a shimmering light, slow pulse light, or very slow pulse from on-to-off depending on the phase of the oscillator frequency.

The LED is in the circuit to verify the presence of positive energy from output pin 4 of IC1. The LED may be in a steady on state because of a very rapid pulse or at a pulse rate that will produce a flicker because of a slow pulse rate (normally below 30 hertz).

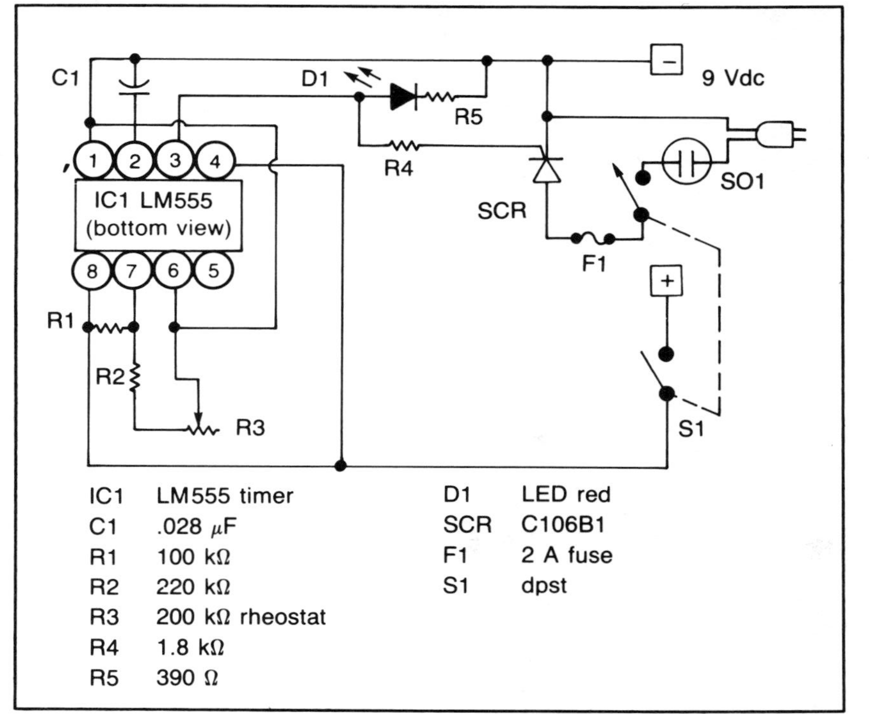

Fig. 9-17. Shimmering light control (display effect).

Further Study

Change C1 slightly upward in value to note an effect of lowering the pulse rate. You can measure the square wave and calculate the time with an oscilloscope. However, remember the scope ground must be at the ground or neutral side of the line. Be careful in working with 120-volts ac because of the lethal potential.

PROJECT 86
VARIABLE POWER CONTROL (DISPLAY LIGHTING)

Function

The effect of an incandescent light blinking at full 120-volt power input is eye catching for display purposes. Colored spotlights like Par 38 lamps and decorative devices like Christmas tree lights can be made to blink at various rates from shimmering to slow pulsations or slow roll-over with the use of this circuit. (See Fig. 9-18.)

IC1, LM555, is operating as a slow pulse oscillator as adjusted by R3 and the C1-C2 combination. One set of pulse changes occur with C1 alone. The addition of C2 by switch S1, lowers the rate to produce slower oscillations and a different effect. The output pin 3 is coupled to the gate of a thyristor via C3, R4 to cross at a frequency close to the 60 hertz ac line or a divisor or multiple of 60 like 10, 20, 30, 90, or 120. Adjusting the R3 rheostat gives the exact frequency. The phase at which the pulse occurs in relationship to the 60 hertz determines the power applied to socket S1 at that time, therefore a variance in power levels and brightness. This circuit is for incandescent lighting only.

Further Study

Design another pulse oscillator with a LM555 that operates at

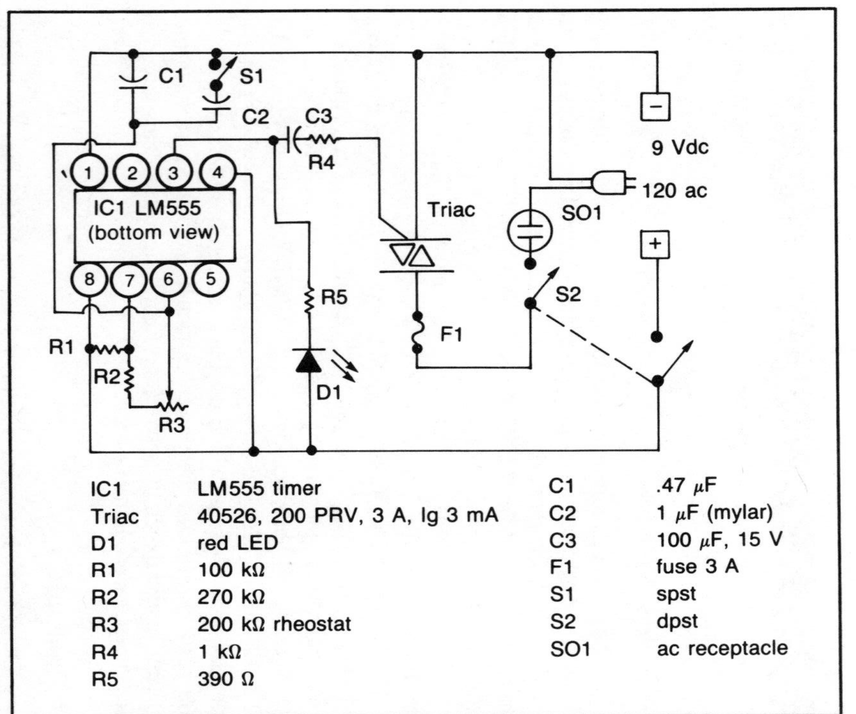

IC1	LM555 timer	C1	.47 μF
Triac	40526, 200 PRV, 3 A, Ig 3 mA	C2	1 μF (mylar)
D1	red LED	C3	100 μF, 15 V
R1	100 kΩ	F1	fuse 3 A
R2	270 kΩ	S1	spst
R3	200 kΩ rheostat	S2	dpst
R4	1 kΩ	SO1	ac receptacle
R5	390 Ω		

Fig. 9-18. Variable power control (display lighting).

a slower pulse rate but couple the output at pin 3 to enabling pin 4 of the existing LM555 to produce a rhythmic lull period when nothing happens. In this manner, the incandescent lamp will appear out and then spring to life only to go out again in a rhythmic pattern.

PROJECT 87
WINDSHIELD WIPER MIST CONTROL

Function

The windshield wiper mist control provides time delay switching to permit the windshield wiper motor to turn on a cycle pattern that is slower than the normal rate. The circuit design is adjustable to provide different time patterns. (See Fig. 9-19.)

IC1, LM386, is a low-voltage audio amplifier IC in a multivibrator circuit design. The output is nonlinear to provide a longer off period than on period—thus a longer delay between wiper sweeps. Pin 3, the noninverting input has a referenced voltage via R1, D1, R2 to output pin 5, which is HIGH when first turned on. The inverting input, pin 2, is connected to C2, R3, which in turn are connected to pin 5. C2 begins to charge because of the on voltage at pin 5. The charge time is controlled by R3, which is adjustable to provide different time periods. When C2 charges to the reference voltage, the output voltage at pin 5 goes LOW and falls to about .5 volts. Capacitor C2 discharges and when it falls below the referenced voltage, pin 5 goes HIGH to repeat the cycle.

The SCR acts as a dc switch, which latches when turned on. However, when the wiper completes the swing, it commutates the SCR and it switches off. Since steady gate current is unnecessary, capaci-

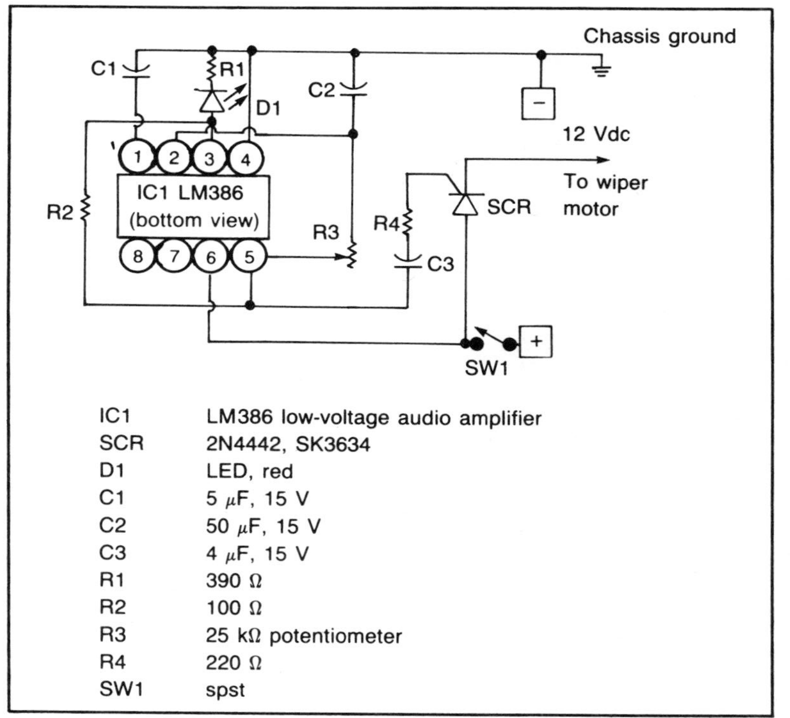

IC1	LM386 low-voltage audio amplifier
SCR	2N4442, SK3634
D1	LED, red
C1	5 μF, 15 V
C2	50 μF, 15 V
C3	4 μF, 15 V
R1	390 Ω
R2	100 Ω
R3	25 kΩ potentiometer
R4	220 Ω
SW1	spst

Fig. 9-19. Windshield wiper mist control.

tor C3 and R4 provides the necessary pulse current to permit completion of the cycle. Resistor R4 might be omitted; however, the .5 threshold voltage at pin 5 might re-gate prematurely, thus it is in the circuit to reduce premature gating. You might find it necessary to increase or decrease R4's value depending upon the SCR used. The LED, D1, is in the circuit also to monitor the on pulse rate.

Further Study

Measure the on/off pulse time at pin 5 with an oscilloscope or even a stop watch. Increase C2's value and not the longer off time. If you use a SCR with a higher gate current, you might have to increase the size of C3. Place a digital multimeter between pin 5 and minus and measure the voltage swing.

PROJECT 88
AUTOMATIC TURN-OFF
POWER CONTROL (ONE HOUR TIME ON-TO-OFF)

Function

A time on-delay-to-off in power control is useful to those who watch television in bed only to fall asleep with the television unattended. This circuit provides a full hour delay before turning off and can be used as an automatic power shut off. (See Fig. 9-20.)

IC 1, CD4081, is a quad 2-input AND gate in which two gates are cascaded to provide two 30 minute delay periods. Gate 1 has pin 1 held HIGH. When SW1 is closed momentarily, C1 charges immediately to the supply voltage. The truth table of an AND gate states that when both inputs are HIGH, the output goes HIGH. The truth condition is met and pin 3 goes HIGH and C2 charges immediately via D1. In gate 2, pin 6 is HIGH and pin 5 goes HIGH because it is connected to C2 which is HIGH. Therefore, output pin 4 goes HIGH to turn on D2, LED monitor, and to provide bias for transistor Q1. Q1 switches on to gate the triac via R5, Q1, and the B+ source. These events happen instantaneously when SW1 is closed.

Upon release of SW1, C2 begins to discharge through R1 and the voltage begins to fall under a long time constant. When C1 discharges to below 4 volts (in about 30 minutes), output pin 3 of

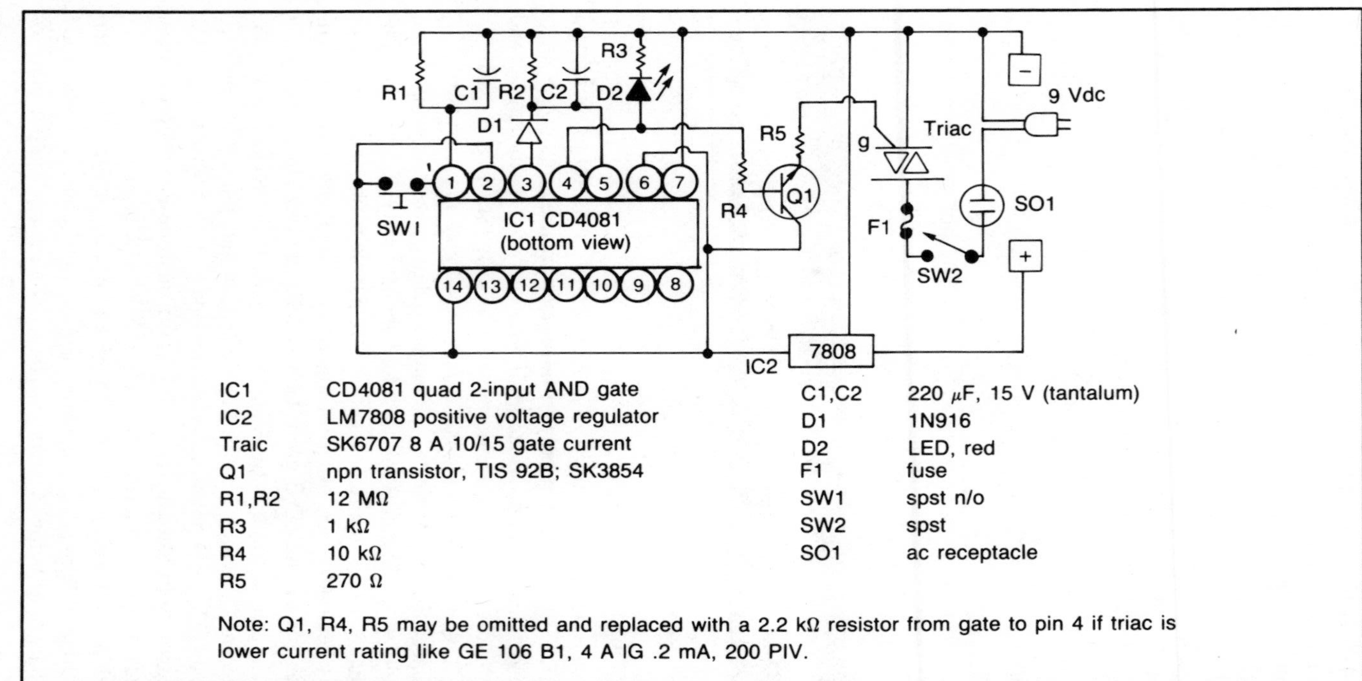

IC1	CD4081 quad 2-input AND gate
IC2	LM7808 positive voltage regulator
Traic	SK6707 8 A 10/15 gate current
Q1	npn transistor, TIS 92B; SK3854
R1,R2	12 MΩ
R3	1 kΩ
R4	10 kΩ
R5	270 Ω

C1,C2	220 μF, 15 V (tantalum)
D1	1N916
D2	LED, red
F1	fuse
SW1	spst n/o
SW2	spst
SO1	ac receptacle

Note: Q1, R4, R5 may be omitted and replaced with a 2.2 kΩ resistor from gate to pin 4 if triac is lower current rating like GE 106 B1, 4 A IG .2 mA, 200 PIV.

Fig. 9-20. Automatic turn-off power control (one hour time-on delay to off).

gate 1 swings low. Capacitor C2 begins to discharge via R2 and it takes about 30 minutes to discharge below 4 volts. This action causes output pin 4 to swing low to reverse bias Q1, which in turn removes the gate current. The triac switches off by ac commutation.

Diode D1 is essential to the circuit as a block to the ground return path of C2. The standby current is 4.9 mA while the operational current is 22.2 mA (of which 13.5 mA is gate current via Q1). IC2, LM7808 voltage regulator is in the circuit to hold the voltage constant and to keep the time accurate.

Further Study

Measure the base and collector currents of Q1 to determine the ratio of I_c/I_b for current gain. There are two other gates that can be added in cascade in the same circuit design as gate 2 to provide two hours of delay time.

PROJECT 89
GARAGE LIGHT CONTROL

Function

An auxiliary light for the garage that turns on when light activated and holds for a specified time is a convenient accessory. The circuit shown here is disabled during the daylight hours and will turn on when triggered by the automobile headlights at night. The time can be set to provide lighting during the period of egress from the car and unlocking the entrance door. (See Fig. 9-21.)

IC1, LM555, is a monostable multivibrator (one shot) that is triggered via LDR2, R2. The LDR is over 1 megohm in darkness, thus the ratio of LDR/R2 sets the trigger voltage HIGH. When light strikes LDR2, it goes LOW to trigger IC1. Output pin 3 goes HIGH to activate the gate of IC2 via R1 and the 100-watt lamp is turned on. The on time is about 4 minutes as determined by C2, R4. Increasing or decreasing these values will change the on time. The trigger action starts the charge of C2 via R4 to the B+. When the charge reaches 2/3 of the applied voltage, the internal flip-flop through pins 7 and 6 causes the output to go LOW to remove the gate voltage and the triac shuts down.

The enabling reset pin 4 must be HIGH in order for the circuit to function. LDR1 should face the exterior daylight in order that the

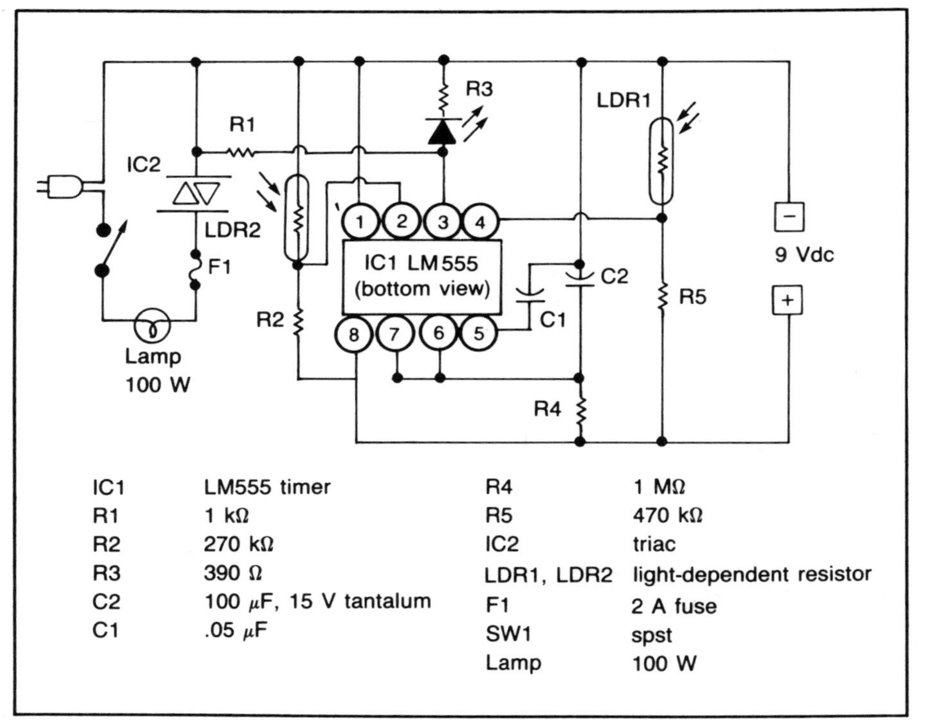

Fig. 9-21. Garage light control.

enable pin 4 is LOW during the daylight hours to disable the IC. The darkness will make LDR1 rise in resistance to change the voltage dividing network LDR1/R5 to make pin 4 to go HIGH and thus enable the circuit.

LDR2 should be in a shroud and located at the back of the garage at headlight height. In this manner the garage doors when raised will permit the headlights to strike LDR2 and turn on the auxiliary light. The shroud can be made of black plastic insulator from an alligator clip. Standby current for the monostable circuit is 5.2 mA and operational current is 19 mA.

Further Study

An iris-like arrangement can be made to cover the LDR1 shroud to set the light limit for triggering. Change C2 to program the time-on period for a particular situation.

PROJECT 90
AC POWER SWITCH
(TIME ON DELAY TO OFF DUAL TRIGGER)

Function

This ac power switch, time-on-delay-to-off, is unique in that it has two methods of turning the circuit on. It would be valuable for the convenience of having a way to switch a circuit on from two locations. (See Fig. 9-22.)

IC1, LM3905, is a precision timer with built-in reference voltage at pin 2 that is regulated internally at 3.15 volts. It remained constant during a swing of supply voltage from 5 to 15 volts. IC1 is utilized as a monostable multivibrator (one shot) with R2-C2 as the components that determine the on time of the on-delay-to-off, which is about 7 seconds. Increasing the value of R2 or C2 will increase the length of the on time. Note that R2 is connected to the 3.15 reference voltage, therefore the on time would be consistent even with a variance in supply voltage. The internal reference voltage of the comparator is 2 volts, therefore C2 will need to charge to about 2 volts to complete the one-shot cycle.

IC1 has an internal transistor in the output stage with a capability of 120 mA, which is more than adequate to gate the triac D2. The operational current was 30 mA under test that was divided three ways;

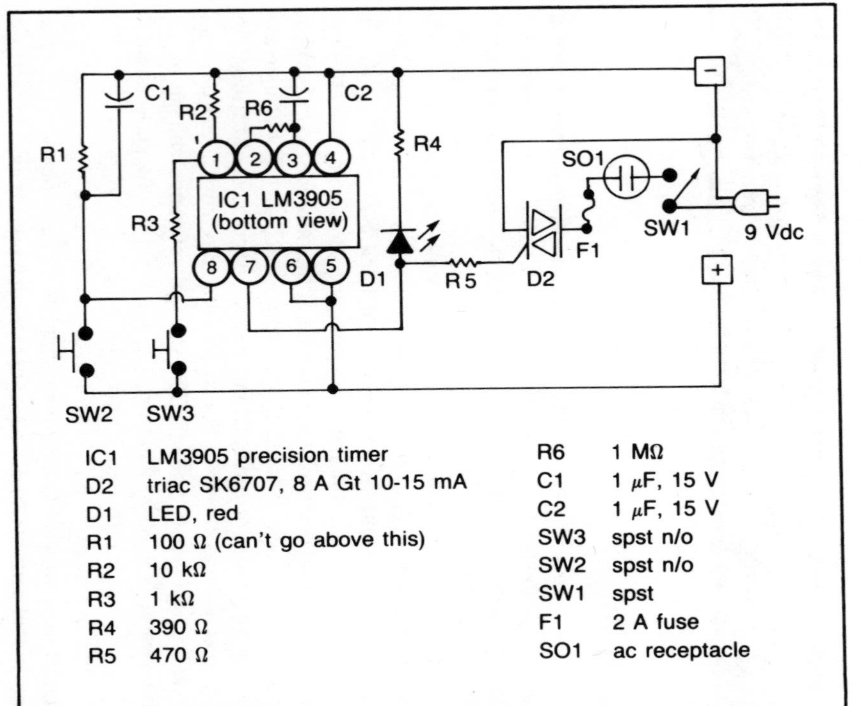

Fig. 9-22. Ac power switch (time on delay to off).

13 mA for the gate current, 15 mA for the LED, and 2 mA in the standby state.

IC1 can be turned on from two locations. SW2 is the logic control via pin 8. When SW2 is closed, a positive voltage is applied for pin 8 and when released the falling voltage provides the negative logic via R1. Also C1 discharges to give a negative kick for positive transition. SW3 is in the trigger circuit of pin 1 with R3 and R2. When S1 is closed, a positive transition pulse at pin 1 triggers on the one shot.

The internal output transistor has the collector pin 6 connected to the supply at pin 5. The emitter at pin 7 provides the current path for D1 (LED) and gate current from the triac.

Further Study

Change the value of C2 to 10 μF, and determine the time to charge to the 2 volts. Can you connect the gate to the collector with appropriate components to get the opposite affect, or light-off-delay-to-on? (Hint: you will need to disconnect R5 from the anode side of the LED; disconnect pin 6 from pin 5 and place an appropriate resistor between pin 6 and the positive source. Connect the unattached end of R5 to pin 6. You will have to change the value of R5.)

PROJECT 91
MAGNETIC TOGGLE SWITCH (AC POWER CONTROL)

Function

A magnetic toggle switch is another way of switching ac power. The input to toggle the IC is a solid-state switch that is activated by a magnet. This switching system prevents bounce because the toggle doesn't take place until the magnet is withdrawn. It would be useful as a hidden switch in which only the knowledgeable person could activate the power on. In this respect it is a safety switching device. (See Fig. 9-23.)

The action is based upon the Hall-effect concept, which states that when a magnet is brought close to an electronic field of a solid-state material, the electrons move away from the magnetic field thus compacting at one side to increase electron mobility. This has the same effect as decreasing resistance and increasing current flow.

IC2, CD4013, is a dual-D flip/flop in which one half (pins 1 through 6) is utilized as a toggle to switch on a transistor, which in turn energizes the gate of a triac. Each toggle turns off-to-on and then on-to-off. The transistor Q1 is necessary because CD4013 can deliver about 2.5 mA at maximum and the triac needs 10 to 15 mA of gate current. IC1 must have set and reset pins 6 and 4 tied to minus.

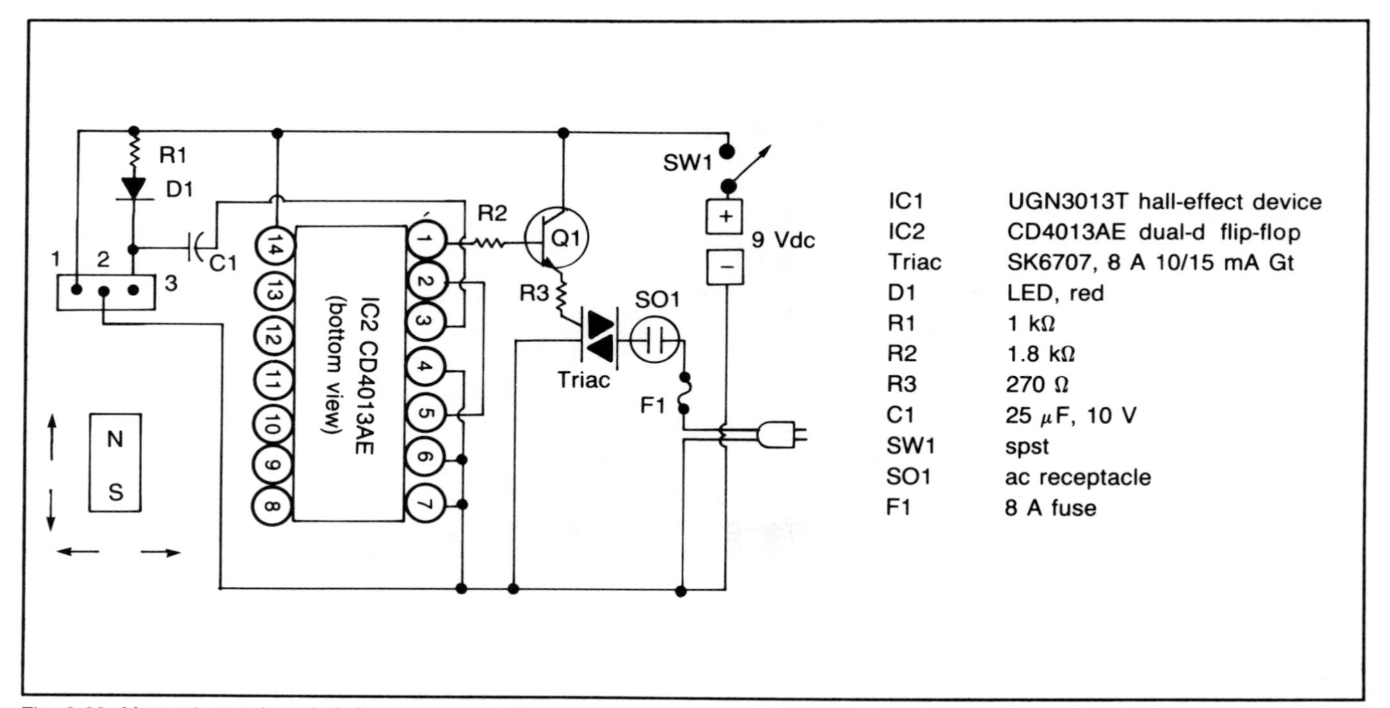

Fig. 9-23. Magnetic toggle switch (ac power control).

Input pin 5 must be connected to $\overline{Q}$ (not) (output pin 2) in order to toggle with each positive transition from the magnetic switch.

IC1, Hall-effect device, when energized via pin 1 and 2 has an output at pin 3 that is HIGH. Since LED D1 is connected to the positive supply via R1 and the cathode end is connected to pin 3 that is positive also, the LED is out. When the north pole of a magnet is brought close to IC1, circle spot side, output pin 3 goes LOW and the LED turns on. This action produces a negative transition at pin 3, which has no effect in the toggling of IC2. However, when the magnet is withdrawn, the device switches and a positive transition occurs at pin 3. The positive transition is applied to the input clock pin 3 of IC2, via C1. IC2 toggles to turn on or turn off the light. The effect of a positive magnet movement prevents bounce and false triggering of the flip/flop. The standby current is 2.6 mA for the circuit, while the operational current is 30 mA.

Further Study

Try the south pole of a magnet to the other side of IC1 and note any change. Also, move the magnet's north pole across the face of IC1 and see if you can turn on the unit. Try different strengths of magnets and determine how far one can be from the face to toggle the circuit.

PROJECT 92
SUPER SWITCH (AC POWER)

Function

The super switch can be used to turn on ac power with a finger touch, air blown from mouth, or some other transducer. It would be useful for a person carrying objects and unable to reach light switch. (See Fig. 9-24.)

IC2, CD4013AE, is a dual-D flip/flop in which one half (pins 1 through 6) is utilized as a toggle to switch on a transistor, which in turn powers the gate to energize the ac socket. Each toggle turns off-to-on and then on-to-off. The transistor, Q2, is necessary because CD4013 can deliver about 2.5 mA at maximum and the triac needs 10 to 14 mA gate current. The bias current for Q2 is .16 mA, which is well within the capability of IC2. IC2 must have set and reset pins 6 and 4 tied to minus. Input pin 5 must be connected to Q (not) output at pin 2 in order to toggle with each positive transition of the clock at pin 3.

IC1, LM555, is designed as a one shot with a 2 second on time, which delivers the positive transition pulse from pin 3 via C3. This time delay prevents a bounce condition because you cannot retrigger another transition until the 2-second delay has transpired. Time-delay time can be changed by changing C1 or R2.

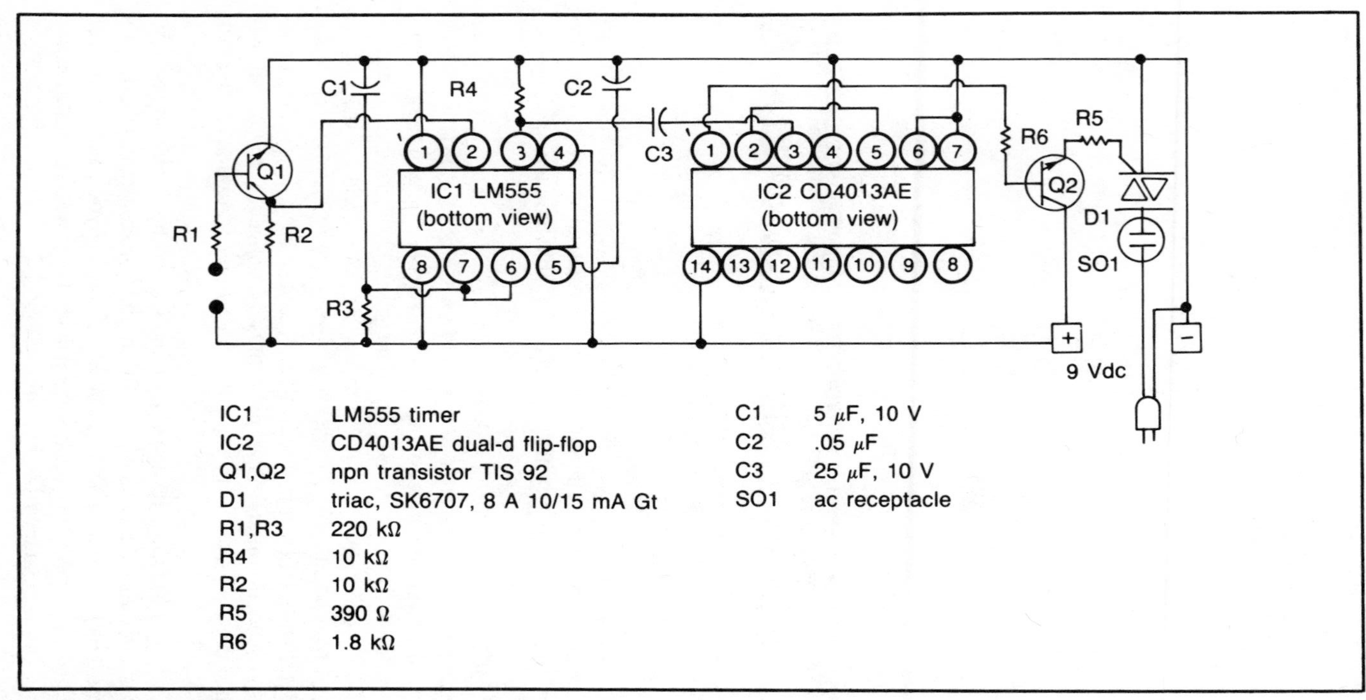

IC1	LM555 timer
IC2	CD4013AE dual-d flip-flop
Q1,Q2	npn transistor TIS 92
D1	triac, SK6707, 8 A 10/15 mA Gt
R1,R3	220 kΩ
R4	10 kΩ
R2	10 kΩ
R5	390 Ω
R6	1.8 kΩ
C1	5 μF, 10 V
C2	.05 μF
C3	25 μF, 10 V
SO1	ac receptacle

Fig. 9-24. Super switch—ac power.

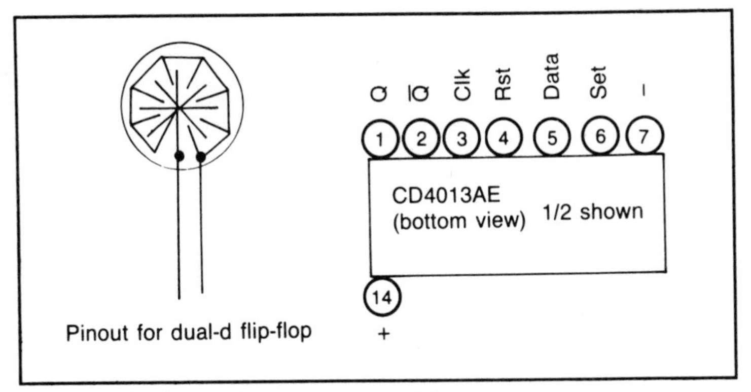

Fig. 9-25. Touch transducer for super switch.

Transistor Q1 provides the trigger pulse to IC1 via collector swing when Q1 is biased on. The touch control as shown in Fig. 9-25 provides positive contact to bias Q1.

Further Study

Try a LDR, light-dependent resistor, as the transducer. Put a shroud around the LDR and then flash a light at the opening to turn the light on and off. Try blowing on the transducer in addition to touching it to turn the light on and off.

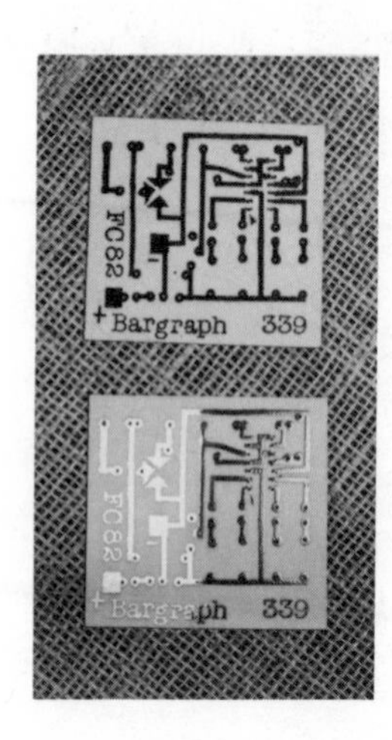

Chapter 10

Testing Control Projects

PROJECT 93
CONTINUITY TESTER (AUDIBLE)

Function

The continuity tester (audible type) shown here has extended possibilities in that the probes are at a square-wave frequency potential of about 500 hertz. This means that one can do continuity checks through some components which is not possible with some other continuity testers. (See Fig. 10-1.)

IC1, ULN2429, is a fluid detector integrated circuit with the audio oscillator part converted to a continuity tester. If the oscillator output frequency at pin 6 is coupled to the internal detector at pin 9, a square wave of 500 hertz appears at the output pins 1 and 14. Both output pin 1 and 14 are utilized to directly drive a 3-inch speaker.

When the probes are touched together, the oscillator frequency from pin 6 is delivered to internal detector pin 9 via C4 and C5, and the oscillator frequency is heard at the speaker. Component C4 could be omitted, but it is added to the circuit to block any dc level that might be available in the event the continuity test was made in a circuit under dynamic conditions. Normally, continuity tests are made in circuits under passive conditions.

Continuity test are utilized to check runs, switches, and some resistive components. In testing this circuit, it was found that circuits

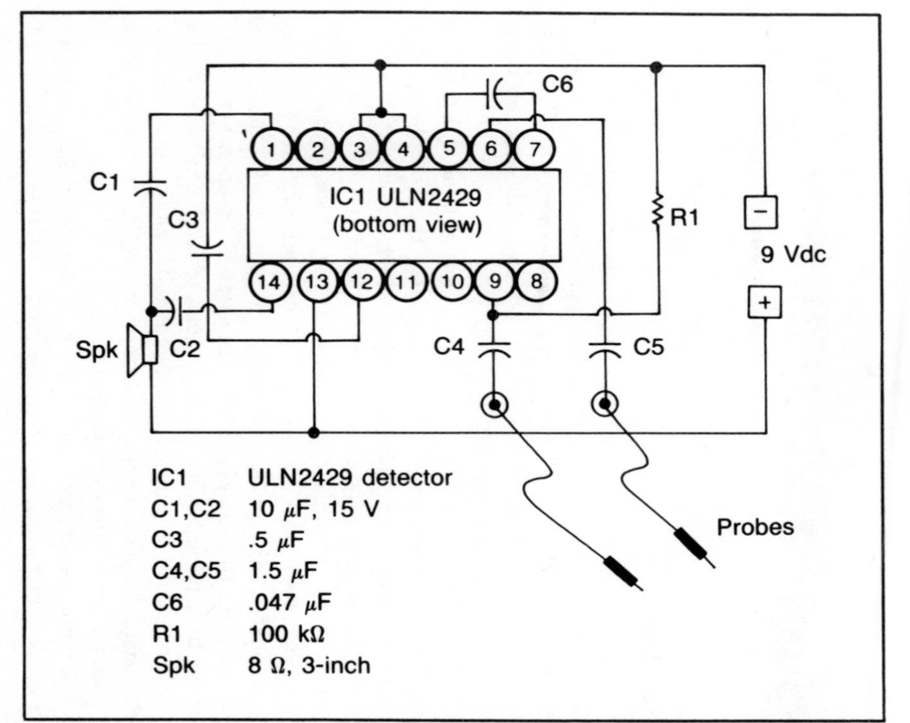

Fig. 10-1. Continuity tester (audible).

could be checked in the passive state through capacitors from .04 to 1500 μF because of the frequency potential at the probes. Likewise, continuity measurement was made through resistors in the range of 2.2 kΩ to 100 kΩ without loss of oscillation.

The standby current is 3.7 mA, which means there is little power loss during idle periods. The operational current is about 20 mA.

Further Study

Try several circuit boards and check continuity of the runs. Try a circuit board with components attached and check runs to include some through capacitors, resistors, switches, and up to the top pins of IC sockets.

Change the value of C6 to .02 and note the change in frequency. You may wish to stay with the .05 because the lower tones are less annoying.

PROJECT 94
CONTINUITY TESTER (FLASHING LIGHT)

Function

The continuity tester is necessary in checking etched circuit boards to insure all copper runs are complete. Hairline cracks, invisible to the eye, can be troublesome if not detected. This tester can measure the continuity of all copper runs including resistors up to 68 kΩ as through connection checks. It can be utilized to check switches, battery connectors, and speaker voice coils for opens. (See Fig. 10-2.)

IC1, quad 2-input NOR gate, has two gates cross-coupled as a slow-speed oscillator of about 10-hertz per second. The feedback capacitor, C1, is open in the feedback path and the oscillation is in the off state. When the probes are shorted, the oscillator starts and pulses are sent to gate 3, pins 9 and 8, via C2. The output pin 10 is coupled to the B- via LED D1, and resistor R3. The LED is on for about 56 milliseconds.

When S1 is closed, the LED will come on because output pin 10 is 4.5 volts HIGH. This LED acts as an on-state indicator in addition to continuity verification when it flashes upon completion of the circuit with the probes. Although the probes are given color leads, there is no particular polarity and the colors can be ignored in usage.

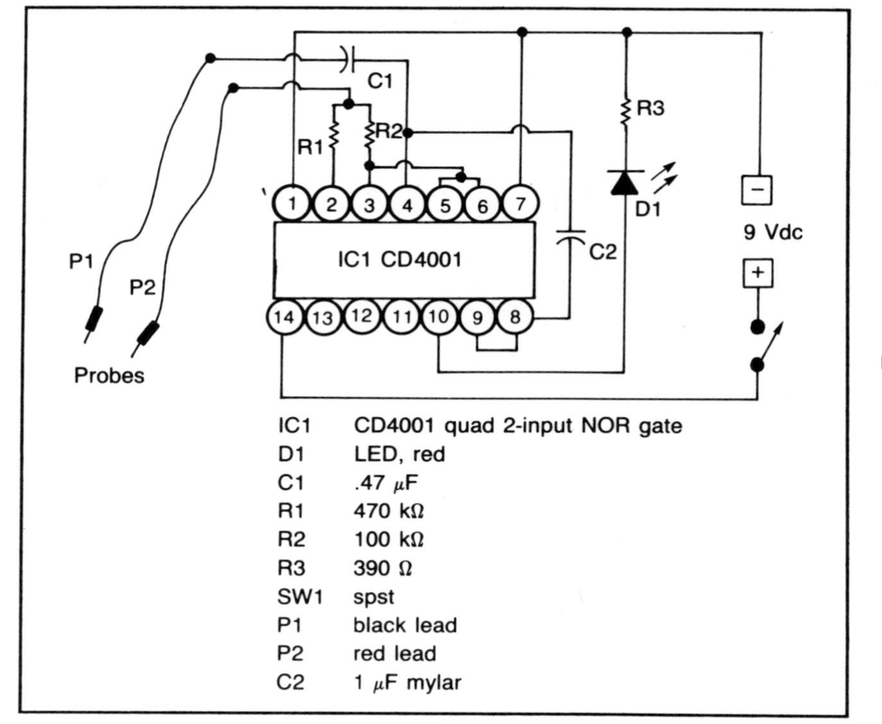

IC1	CD4001 quad 2-input NOR gate
D1	LED, red
C1	.47 μF
R1	470 kΩ
R2	100 kΩ
R3	390 Ω
SW1	spst
P1	black lead
P2	red lead
C2	1 μF mylar

Fig. 10-2. Continuity tester.

Further Study

Try placing a .47 μF capacitor in parallel with C1 and note the change in the pulse flash rate when making continuity tests.

PROJECT 95
TIMER TESTER

Function

The timer integrated circuit such as the LM555 or LM7555 are so valuable in design applications that a tester is indispensable. Misconnections on these ICs usually means overheating and damage that can cause nonfunction. Sometimes faulty new ICs are missed in inspection and are delivered inoperative. This circuit design is unique in that it tests all elements. (See Fig. 10-3.)

IC SO1 is the socket for insertion of the LM555 to be tested. It is designed to make the timer function as an astable multivibrator of about 4-hertz per second to produce a nearly symmetrical square wave. The output pin 3 is coupled to two LEDs via resistors to indicate both the positive and negative swing in the wave. The green LED is in the positive supply leg to light when pin 3 swings LOW and the red LED is in the negative leg of the supply to light when pin 3 swings HIGH. A spdt switch, SW1, is in the circuit to connect RESET pin 4 either HIGH or LOW. When set LOW to minus, the output at pin 3 is LOW and the green LED is on. In this state the 555 is disabled. When RESET is high the oscillation is enabled and the green-red LEDs flash alternately. The final test is the control voltage at pin 5.

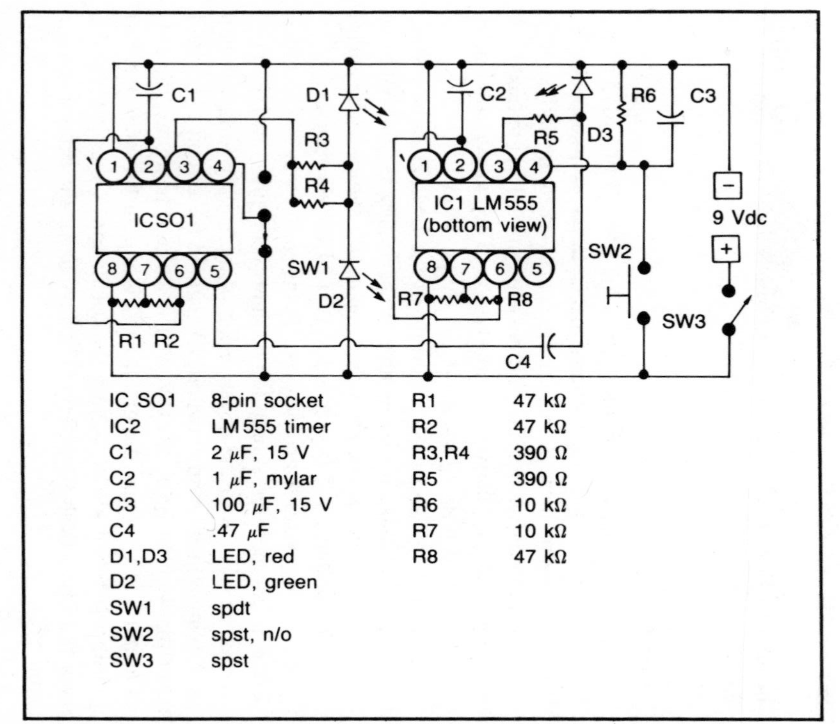

IC SO1	8-pin socket	R1	47 kΩ
IC2	LM555 timer	R2	47 kΩ
C1	2 μF, 15 V	R3,R4	390 Ω
C2	1 μF, mylar	R5	390 Ω
C3	100 μF, 15 V	R6	10 kΩ
C4	.47 μF	R7	10 kΩ
D1,D3	LED, red	R8	47 kΩ
D2	LED, green		
SW1	spdt		
SW2	spst, n/o		
SW3	spst		

Fig. 10-3. Timer tester.

IC2, LM555, is another astable multivibrator at 14-hertz per second as determined by C2, R7, and R8. The circuit is designed to make the LM555 a burst oscillator at about 3-seconds duration as determined by C3 and R6. When SW2 is momentarily closed, C3 charges to the supply voltage and RESET pin 4 is HIGH to enable IC2 to oscillate as indicated by LED D3. Upon release of SW2, capacitor C3 discharges through R6 and produces a falling voltage. During the time that C3 discharges from full voltage to about .9 V, the oscillation is maintained. The output pin 3 is coupled via C4 to pin 5 of the 555 under test. This controlled-oscillation voltage at pin 5 causes the output pin 3 of the 555 under test to oscillate at the same frequency as the burst (or all three LEDs flash at the 14-hertz rate).

What is being tested is the ability of the 555 to produce a swing voltage, to be disabled by RESET, and/or to be controlled via pin 5. The IC is good if all these conditions are met.

Further Study

Note the output of the IC being tested is nearly symmetrical and thus the green and red LEDs are on and off in equal increments. In testing a number of LM555 ICs, it was noticed that some were found that did not produce the full swing to indicate 100 percent. In other words, the green LED did not go completely out in the same manner as the red LED, and thus indicated less than perfect performance. Try several LM555s to verify performance.

PROJECT 96
SCR/TRIAC TESTER

Function

An SCR/Triac Tester for good/bad condition and type identification can be made with very few parts. This is a particularly useful tester because purchasing bulk quantities of these components often means that there is no number identification or perhaps there are even mixed types. The savings of this kind of purchasing are great, but one must be able to sort and identify. (See Fig. 10-4.)

IC1, LM555, is in the circuit as a monostable oscillator with an on time of about 1.5 seconds, which is enough time to make a test and not worry about over-gating. The time of the monostable (one shot) is determined by C1, R2. Pressing SW1 (n/o switch) will trigger the one shot and output pin 3 goes HIGH. Monitor LED D1 comes on and voltage is applied to SW2. SW2 switches in R4 (low-gate current control) or R5 (high-gate current control) to the gate pin of the socket. When the ac plug is energized, Lamp 1 comes to full brightness and becomes the reference power. The gate current when SW1 is switched will turn on the SCR or Triac and lamp 2 will come on.

If the gate current is sufficient and the unit under test is an SCR, lamp 2 should be at half brightness to compare with the reference lamp 1. (Both lamps should be of the same wattage.) If a triac is under

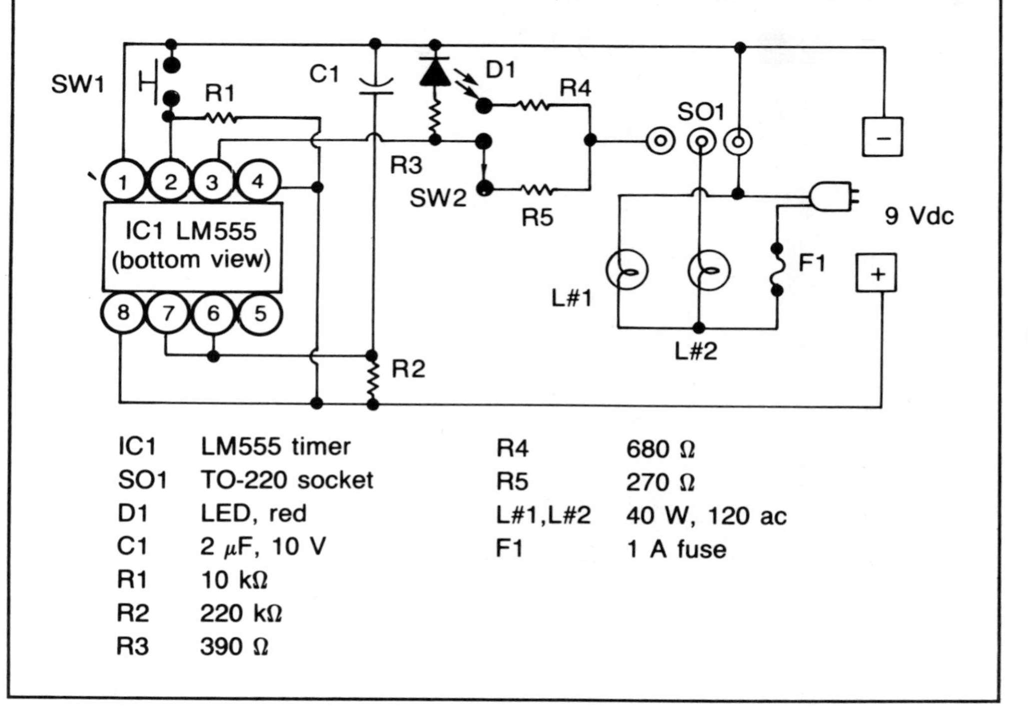

Fig. 10-4. SCR/triac tester.

test, the lamp should be at full brightness and again compared with the reference lamp. The resistor R4 will produce a low-gate current of 10 mA and resistor R5 will produce a high-gate current of 25 mA, which should measure most SCRs and triacs.

You may have to switch to high-gate current for some SCRs and triacs otherwise some triacs will be at half brightness due to lack of sufficient gate current. The 1.5 second short-look cycle would not be on long enough to cause gate current damage.

If lamp 2 should come on with no gate current, then the SCR or triac is shorted. If the SCR or triac cannot be gated in either switched position, it is probably open.

If the triac or SCR is numbered, then you should get the specifications from technical manuals as to gate current. This tester cannot check for PIV, therefore you should also check technical manuals for this. Remember, a 120-volt ac line has a PIV of about 170 volts; this tester should be used with SCRs or triacs that can meet this requirement before testing. It is possible if one has a variac to lower the ac rms to the tester and run tests with SCRs and triacs with lower PIVs.

Further Study

Open R4 and R5 and measure the gate current with the component under test to verify the gate current. Note a 1000-ohm resistor with a 9-volt supply will permit a 6-mA gate current and a 220-ohm resistor will permit 33 mA of gate current.

PROJECT 97
GATE TESTER

Function

The quad 2-input gates of OR, NOR, AND, and NAND are excellent integrated circuits for delay timing, switching, oscillating, and other control interfacing. Any one or all four gates can be utilized in particular applications. The loss of any one gate doesn't mean necessarily the discarding of the whole chip. The tester shown here is an easy way to check the condition of each gate. (See Fig. 10-5.)

IC SO1 is a 14-pin socket for gate insertion in testing. Be sure to observe the pin 1 position when inserting the gates (the bottom view mark on the schematic indicates position 1). The circuit is designed to check all four gates simultaneously in accordance with their truth tables. The inputs A and B to each gate are switched HIGH or LOW via SW1 and SW2 and the output is read with a LED for each gate.

For example, pins 1, 6, 13, and 8 are A inputs for each gate that are switched high or low via SW1. Likewise, pins 2, 5, 12, and 9 are B inputs for each gate that are switched high or low via SW2. The LEDs for corresponding gates are located next to each gate, therefore it is easy to determine which of the outputs are turned on.

You must switch the gates HIGH or LOW according to the truth

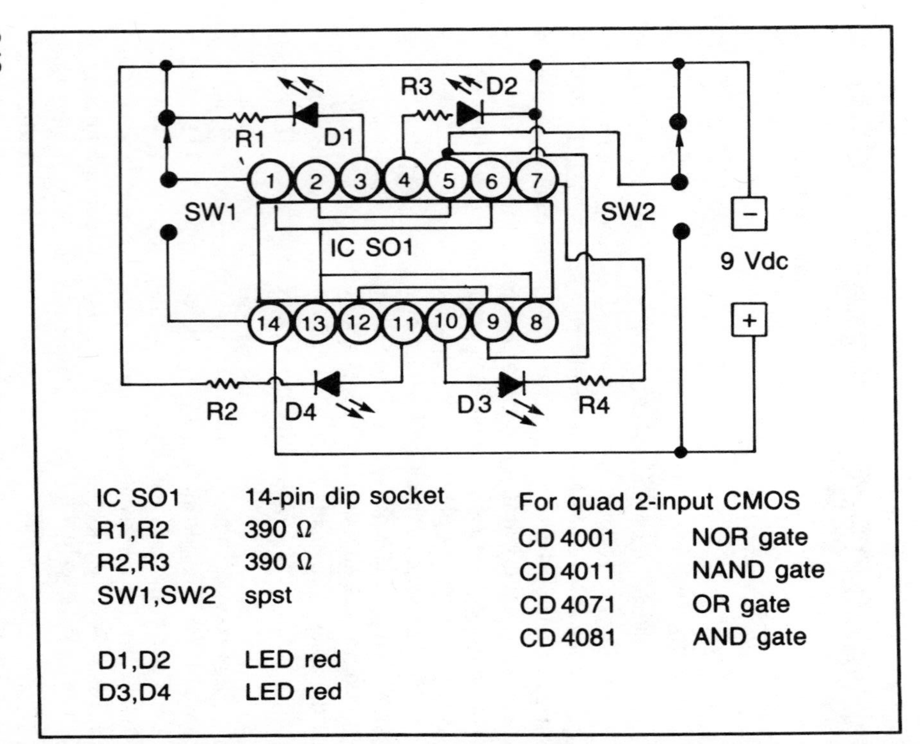

Fig. 10-5. Gate tester (CMOS).

Table 10-1. Truth Table for Each Gate that Identifies What the Output will be in Relation to the Inputs.

CD4071 OR Gate

A	B	Out
L	L	L
L	H	H
H	L	H
H	H	H

CD4001 NOR Gate

A	B	Out
L	L	H
L	H	L
H	L	L
H	H	L

CD4081 AND Gate

A	B	Out
L	L	L
L	H	L
H	L	L
H	H	H

CD4011 NAND Gate

A	B	Out
L	H	H
L	L	H
H	L	H
H	H	L

tables (Table 10-1) to determine if the outputs meet the condition of going HIGH as indicated by lighted LEDs.

Further Study

Try several OR, NOR, AND, and NAND gates against the truth tables to check if the conditions can be met.

PROJECT 98
STEREO AMPLIFIER TESTER

Function

It is often necessary to measure and compare the performance of a stereo amplifier to be sure that the balance control, volume linearity, equality of stage gain, right- and left-channel differential, crosstalk, and other factors are meeting expectations. This Stereo Amplifier Tester measures these conditions and is useful as a troubleshooting device. (See Fig. 10-6.)

IC1, LM556, is a dual audio generator with a symmetrical-square wave output as determined by R1, R2, and R3, R4. The frequency is approximately 1000 and 500 hertz via SW1 and SW2. The 1000 hertz was selected as base because specifications are based on this frequency. The designer can establish other frequencies by changing the values of C1, C2, and C3, C4. The output is coupled via C5 and C6 to separate level controls for adjusting balance or to favor either channel by choice.

The outputs are coupled to IC2, LM339 quad comparator, that utilizes two comparators for each channel as an inexpensive method of metering the drive levels. The reference voltages for each comparator are set by R13, R12, R11, and R15, R16, and R17. These are adjustable via rheostat R11, and R17. The differential caused by

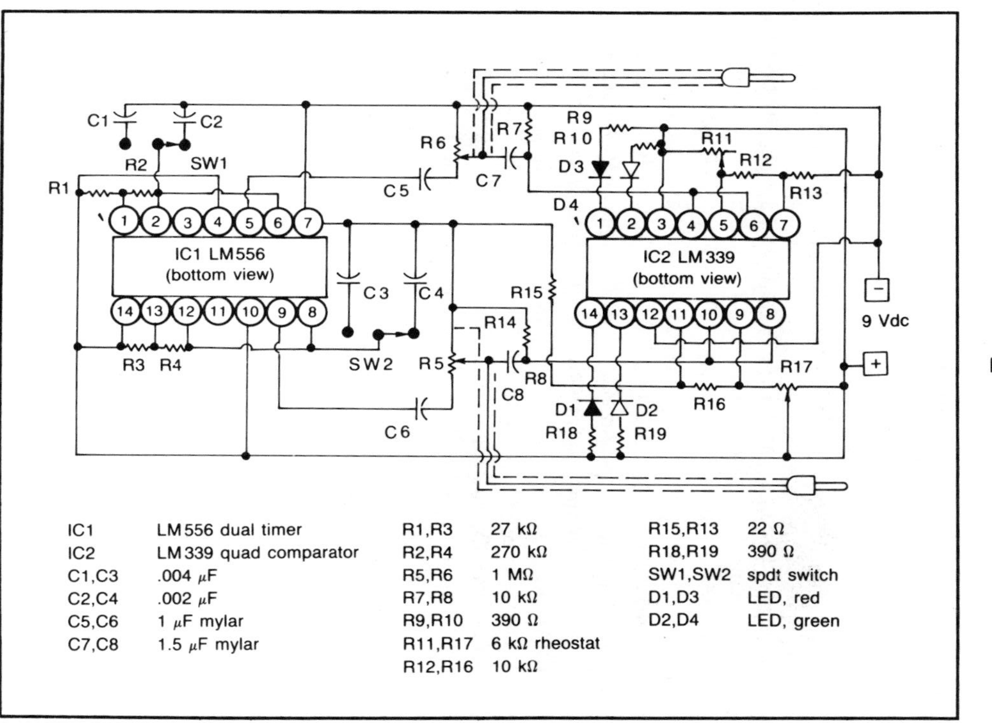

Fig. 10-6. Stereo amplifier tester.

the R13 (22 Ω) and R12 (10 kΩ) resistors was planned to turn on the first and second LEDs at different levels. This feature is essential to know the P-P drive levels when each LED comes on. The first LED comes on at between 20 and 70 mV, while the second LED comes on at 2 V P-P. This gives two input levels for preamplifier and tuner inputs respectively. You may wish to turn the LED on and then adjust drive until it just goes off, thus getting a more precise point. A two LED system for input drive balance or differential requires interpolation of frequency between the two lights. If R5 and R6 are equal, one could establish exact dial settings that record specific drive levels.

Another option would be to place a digital voltmeter across R10 and R18 and adjust the drive to set the levels, or to use a dual-trace oscilloscope to measure the P-P voltages in a more precise manner or to calibrate a dial setting.

Further Study

How about utilizing two LM339 comparators to provide four LEDs per channel as a metering system. Remember, you can set R13, plus two other resistors to provide four fixed P-P points. The possibility of overdrive is very real, particularly on preamplifier inputs; therefore you may wish to set the four LEDs in the metering system to read between 4 and 100 mV P-P. Remember, the resistors can be non-equal to provide selected spacing, like 10, 22, 220, 470, and rheostat.

You can also provide for more preset frequencies by switching additional capacitors like C1 and C2. These capacitors could be plug-in types that can be changed to accommodate particular frequencies.

PROJECT 99
LOGIC TESTER

Function

The logic tester is useful in checking gates, clocks, other pulse devices to determine voltage logic at the input and output pins. It can trace those pins to the board for continuity checks. (See Fig. 10-7.)

IC1 and IC2 are BiFET-input operational amplifiers. They are designed as parallel comparators with inverting and noninverting input pairs tied together. Pin 2 has a reference voltage via R2, R4. Pin 3 has a reference via R1 and R3, which is set the same as pin 2 or 2.14 volts from common. IC1 has an offset adjustment via R5 and it is set to reverse-bias pin 6 to about +7 V. Thus, D1 (green LED) is off. IC2 has an offset adjusted via R6 to reverse-bias pin 6 to about −6 V. Thus, D2 (red LED) is off.

A bipolar supply or two 9-volt batteries are utilized to power the circuit. This makes the logic tester independent of the voltages of the circuit under test. The logic circuit could be kept in better balance by utilizing a 7808 positive regulator and 7908 negative regulator in each of the bipolar legs. This would compensate for battery wear.

The circuit is used with the common tied to the ground or common of the circuit under test. The comparator input pin 3 is the probe. If a HIGH or positive voltage is found, D2 (red LED) turns on to in-

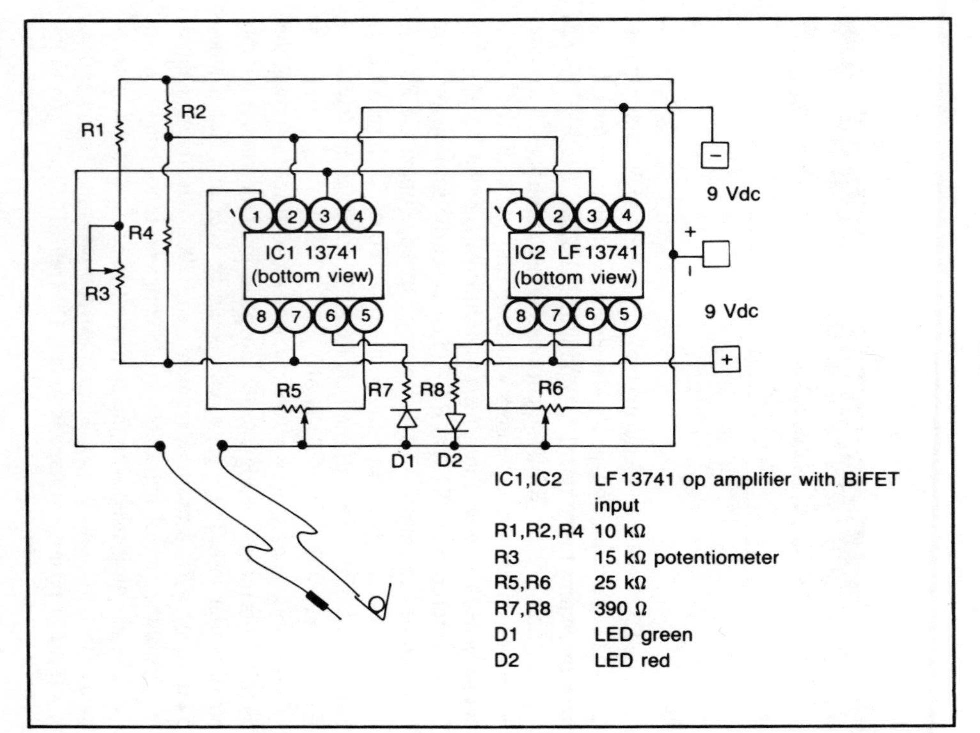

Fig. 10-7. Logic tester.

dicate positive logic. If a LOW is found, D1 (green LED) turns on to indicate negative logic. If both LEDs flash alternately, the logic is a switching pulse.

Further Study

Measure the offset voltages at pin 6 for each comparator. Secure a pulse circuit like a LM555 oscillator or a gate circuit and measure the logic.

PROJECT 100
OPERATIONAL AMPLIFIER TESTER

Function

A test for an operational amplifier is to get the output to swing to the negative and positive voltages of the supply. This can be accomplished by a circuit that makes the operational amplifier to function as a multivibrator. The tester described here will check (8 pin dip) Op Amps such as u741, LF351, TL081, and LM308 variety. (See Fig. 10-8.)

When SW1 (switch) is closed, the output pin 6 is HIGH and capacitor C1 begins to charge via R2. Also, LED D1 is on. Components R1 and R3 to pin 6 provide a voltage divider that sets a reference voltage at pin 3. When C1 reaches this reference level at pin 2, the output goes LOW and LED D2 conducts while LED D1 goes off. Capacitor C1 discharges through the on LED and R2, thus lowering the voltage at pin 2 to begin the next cycle. The output is almost symmetrical so that the on time of each LED is about equal. The total time each light is about 140 milliseconds or seven alternate flashes per second. Changing C1 to a larger value will slow the flash rate.

If the output doesn't swing to flash each light alternately, the IC is no good. Be sure the LEDs are in correctly. Remember, the flat spot on the side is the cathode. If they are placed incorrectly, both LEDs will flash at the same time. The batteries must be equal in volt-

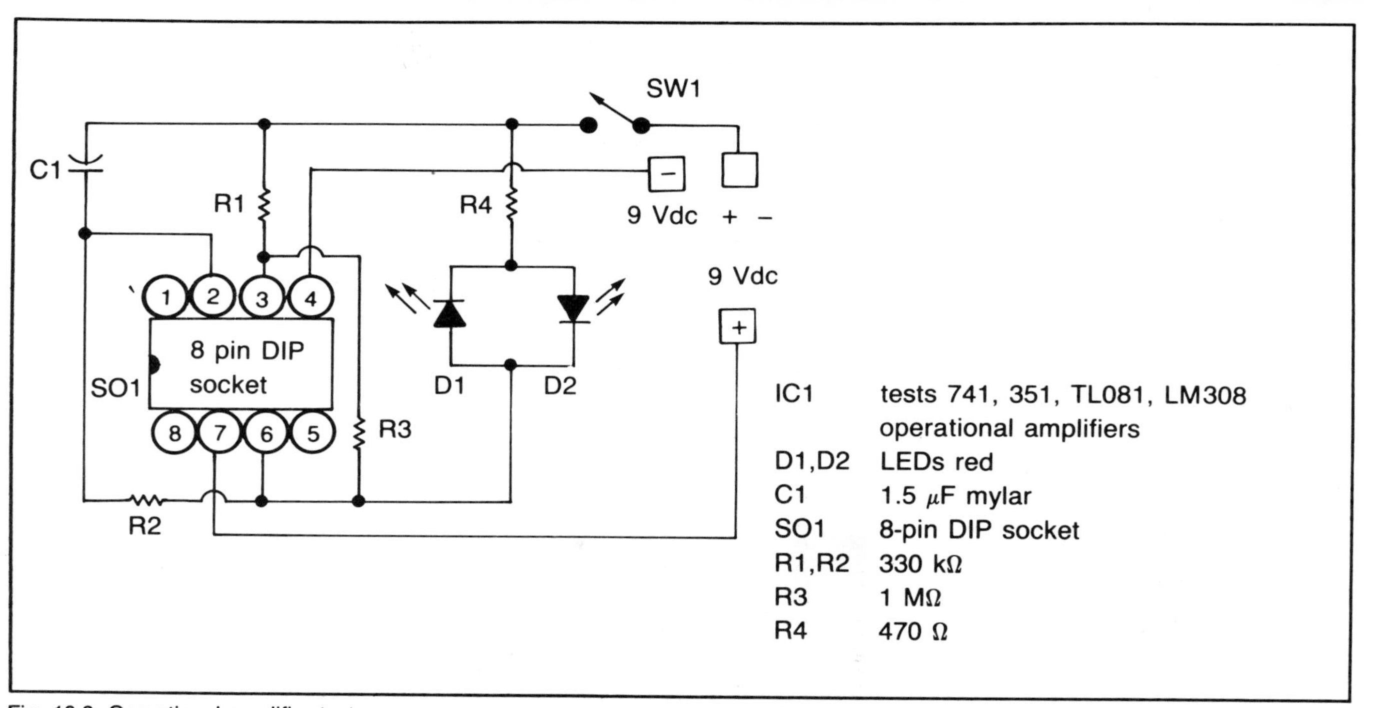

Fig. 10-8. Operational amplifier tester.

age level or else one LED will flash brighter than the other.

Further Study

Change the value of C1 to 1 μF and note the change in the flash rate. Also, change the value of R2 to note the change in the flash rate. Try two batteries of different voltages like 9 V and 7 V and note the difference in intensity of the LEDs. Reverse one of the diodes and note the difference.

PROJECT 101
DIODE/TRANSISTOR TESTER

Function

The diode/transistor tester is an inexpensive circuit for determining the relative merit of light-emitting diodes, power and signal diodes, and transistors. The circuit will test the good or bad state of these devices. A more sophisticated analyzer is necessary to measure beta, and leakage such as I_{ceo} or I_{bco}. (See Fig. 10-9.)

IC1, TL081, is an operational amplifier in a multivibrator design to provide a positive and negative voltage swing at the output of pin 6. The offset balance of the op amp is set to zero by R2 and pins 1 and 5. This is accomplished by removing C1 to stop oscillation and by measuring the voltage between pin 6 and common with a digital multimeter while adjusting R2 to set the reading to near zero.

The swing voltage is applied to the collector pin of the transistor socket TSO1 via two LEDs in reverse configuration; the emitter pin is connected to common. The base pin is connected also to pin 6 via R5 to provide bias voltage for the transistor under test. Also a crystal-type socket DSO1 is connected across the emitter-collector junction of TSO1 for diode insertion and measurement.

If a pnp transistor is inserted into the socket TSO1 and pin 6 is in the negative voltage swing, current will flow through R5 to base-emitter direction to bias on the pnp type. Also, pin 6 applies negative

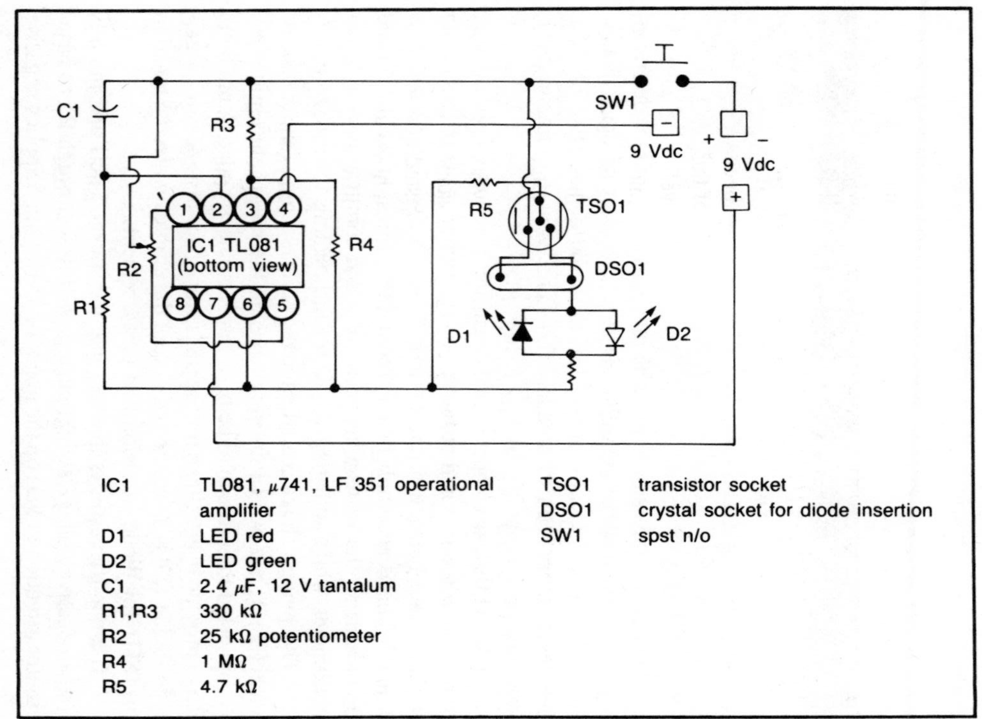

Fig. 10-9. Diode/transistor tester.

voltage to the LED D2 in collector-emitter direction. Thus, the green LED will flash to indicate the transistor is conducting properly and is in good condition. This green LED will continue to flash on negative swings as long as SO1 is closed.

If a npn transistor is inserted into socket TSO1 and pin 6 of the multivibrator is on the positive voltage swing, current will flow from common to emitter-base via R5 to bias on the npn type. Also, pin 6 applies a positive voltage to the LED D1 to cause current to flow from emitter to collector and through the red LED to cause it to flash, thus indicating it is conducting or in good condition. In both pnp and npn testing, there will be no flashing if the transistor is open or both will flash if transistor is shorted. Identifying the transistor family as npn or pnp (in addition to its condition) is a first step in testing component usability.

A LED can be checked with this tester. Place the LED in the crystal socket without concern for polarity. This completes an emitter to collector connection and a good LED will flash when proper polarity voltage appears. If the green LED flashes, the cathode side of the LED is toward the anode of the green LED. If the LED is turned around and reinserted, the red LED will flash because the cathode side of the LED is away from the cathode side of the LED. This test helps to identify the condition of the LED as well as to determine which leg is the cathode side. If the LED is open, no LEDs will flash. If the LED is shorted, both the green and red LEDs will flash.

A signal, or power diode can be tested on this unit in the same manner as the LEDs. The cathode end (which is often unmarked) can be determined.

Power transistors can also be tested, however a special socket adaptor will be needed to insert the leads into a standard signal transistor socket. Several power transistors were tested and those with high leakage made both green and red LEDs flash; however, one LED flashed brighter then the other to indicate a gain differential even with leakage.

A bipolar converted supply would be necessary to make the unit ac energy input. Two 9-volt transistor batteries should provide excellent service. Since turn on is a spst n/o switch, the standby energy is zero.

Further Study

Try several LEDs, power diodes, and transistors to become familiar with the operation. Secure both good and bad types to check your reading of the condition.

Appendices

Appendix A
List of Integrated Circuits Used in The Text

CD4001	Quad 2-Input NOR Gate
CD4009	Hex Inverter
CD4011	Quad 2-Input NAND Gate
CD4013	Dual-D Flip/Flop
CD4016	Quad Bilateral Switch
CD4071	Quad 2-Input OR Gate
CD4081	Quad 2-Input AND Gate
LF351	High-Performance Operational Amplifier
LF353	Dual-FET Operational Amplifier
LF13741	Operational Amplifier BiFET Input
LM308	Super Operational Amplifier
LM311	High-Performance Comparator
LM339	Quad Comparator
LM377	Dual 2-Watt Amplifier
LM378	Dual 4-Watt Amplifier
LM384	5-Watt Audio-Power Amplifier
LM386	Low-Voltage Amplifier
LM555	Timer
LM556	Dual Timer
LM714	Operational Amplifier
LM3905	Precision Timer
LM3909	Flasher Oscillator

LM3914	Dot/Bar Display Driver
MC1458	Dual Operational Amplifier
MC14017	Divide-By-10 Counter/Decoder
MC14027	Dual-JK Flip/Flop
MC14046	Phase-Locked Loop
MC14072	Dual 4-Input OR Gate
MC14073	Triple 3-Input AND Gate
UGN3013T	Hall-Effect Device
ULN2429A	Fluid Detector
ULN3330Y	Optoelectronic Switch
ULN3701	Power Amplifier
TIL113	Photo-Darlington Opto Coupler
TL081	Operational Amplifier
ZN414	Ferranti AM IC

Appendix B
List of Electronic Suppliers

Many of the components for constructing the circuits in this text can be purchased from the local electronic supplier. The following mail-order supply houses are listed for those who choose this method as the source of supplies.

Parts-Components-Equipment-Hardware-Plastics-Chemicals

Advance Process Supply Co.
Skillman Ave, @ 34th Street
Long Island City, NY 11101
(CDF2 and CDF4 Presensitized Screen process)

Circuit Specialists Inc.
PO Box. 3047
Scottsdale, AZ 85257
(ICs, transistors, diodes, LEDs and other components.
Good catalog with Base Diagrams).

Digi-Key Corp. PO Box 677
Highway 32 South
Thief River Falls, MN 56701
(ICs and Components)

ETCO Electronics
Rt 9 North
Plattsburgh, NY 12901
(Assorted Supplies/Positive Photo Sensitizer in Spray Can)

Jameco Electronics
1021 Howard Avenue
San Carlos, California, 94070
(All kinds of ICs)

Kelvin Electronics Inc.
1900 New Highway
Farmingdale, NY 11735
(Etching Chemicals and Supplies)

Kepro Circuit Systems Inc.
3630 Scarlet Oak Blvd.
St. Louis, MO 63122
(Etched Circuit Equipment and Supplies)

References

Fairchild Semiconductor, *MOS/CCD Data Book*. Fairchild Semiconductor Corporation, Mountain View, CA.

____________, *Linear Integrated Circuits Data Book*. Fairchild Semiconductor Corporation, Mountain View, CA.

General Electric Company, *Replacement Semiconductor Guide*. General Electric Company, Owensboro, KY.

Grossblatt, Robert. "Finishing Up About Regulators." Radio-Electronics, (August 1983), page 98 ff.

____________, "Voltage Regulators and Power Supplies." Radio-Electronics,(May 1983), page 90 ff.

Intel, *Peripheral Design Handbook*. Intel Corporation, Santa Clara, CA.

Jung, Walter G., *IC Op Amp Cookbook*. Indianapolis, Indiana: Howard Sams Company, 1976.

Marston, Ray. "All About Power Supply Circuits." Radio-Electronics. (July 1984), page 65 ff.

Motorola Inc., *Linear And Interface Integrated Circuits*. Motorola Semiconductor Products, Inc., Phoenix, AZ.

____________, *Voltage Regulator Handbook, Theory and Practice*, Motorola Semiconductor Products, Inc., Phoenix, AZ, 1976, 202 pages.

National Semiconductor, *Audio/Radio Handbook*. National Semiconductor Corporation, Santa Clara, CA.

_____________, *Voltage Regulator Handbook.*
_____________, *CMOS Integrated Circuits.*
_____________, *Special Function Data Book.*
_____________, *Linear Data Book.*

Radio Corporation of America, *RCA SK Series.* RCA Solid State Division, Somerville, NJ., 1983.

_____________, *COS/MOS Digital Integrated Circuits,* Data Book Series.

Tritely, Harry L. "All About Thermistors." Radio-Electronics, (January 1985), page 47 ff.

Index

Index

Other Bestsellers From TAB

☐ **BASIC INTEGRATED CIRCUITS—Marks**

With the step-by-step guidance provided by electronics design expert Myles Marks, anyone with basic electronics know-how can design and build almost any type of modern IC device for almost any application imaginable! In fact, this is the ideal sourcebook for every hobbyist or experimenter who wants to expand his digital electronics knowledge and begin exploring the fascinating new possibilities offered by today's IC technology. 432 pp., 319 illus.
Paper $16.95 Hard $26.95
Book No. 2609

☐ **THE 555 IC PROJECT BOOK—Traister**

Now, with this exceptionally well-documented project guide, you can begin to use the 555 timer IC in a variety of useful projects that are both easy and fun to put together. From an IC metronome to a ten-second timer and even an electronic organ, this is an exciting collection of 33 different projects that even a novice at electronics can make easily, inexpensively, and in only a few hours time. Most important it provides the know-how to start you designing your own original 555 IC projects to suit your own special needs. 224 pp., 110 illus.
Paper $11.95 Hard $18.95
Book No. 1996

☐ **55 EASY-TO-BUILD ELECTRONIC PROJECTS**

If you're looking for a gold mine of exciting, fun-to-build, *and useful* electronics projects that are both easy and inexpensive to put together . . . here's where you'll hit pay dirt!! Here are more than 50 unique electronic projects selected from the best that have been published in *Elementary Electronics* Magazine. . . projects that have been tested and proven! 256 pp., 256 illus. 7″ × 10″.
Paper $14.95 Hard $21.95
Book No. 1999

☐ **THE ENCYCLOPEDIA OF ELECTRONIC CIRCUITS—Graf**

Here's the electronics hobbyist's and technician's dream treasury of analog and digital circuits—nearly 100 circuit categories . . . over 1,200 individual circuits designed for long-lasting applications potential. Adding even more to the value of this resource is the exhaustively thorough index which gives you instant access to exactly the circuits you need each and every time! 768 pp., 1,782 illus. 7″ × 10″.
Paper $29.95 Book No. 1938